Psychological and Neurobiological Aspects of Eating Disorders

Nathalie Tatjana Burkert

Psychological and Neurobiological Aspects of Eating Disorders

A Taste-fMRI Study in Patients Suffering from Anorexia Nervosa

Springer

Nathalie Tatjana Burkert
Graz, Österreich

Zugl. Dissertation, Medizinische Universität Graz 2015

OnlinePlus material to this book can be available on
http://www.springer.de/978-3-658-13067-1

ISBN 978-3-658-13067-1 ISBN 978-3-658-13068-8 (eBook)
DOI 10.1007/978-3-658-13068-8

Library of Congress Control Number: 2016935721

Printed on acid-free paper

This Springer imprint is published by Springer Nature
The registered company is Springer Fachmedien Wiesbaden GmbH

To the most important individuals in my life –
Albert, Frederick and Fluffy.
Thank you for your endless love and support.
I love you more than anything.
You'll be in my heart forever.

Acknowledgements

I dearly want to thank many people who have supported me during the last four years as I conducted this study and wrote this thesis.

First of all, I want to thank my supervisor, principal, and mentor Univ.-Prof. Dr. Wolfgang Freidl who made it possible in the first place to conduct this study. He supported me during the past few years with his broad knowledge, encouraged me to do research in my fields of interest and always gave me advice concerning methodological as well as theoretical questions. His invaluable support throughout my career has enabled me to become the researcher I am today. I am incredibly thankful to him.

I also want to thank Univ.-Doz. Dr. Elfriede Greimel and Univ.-Prof. Dr. Peter Stix who supported and supervised me while doing this research in the field of eating disorders.

I especially want to thank Mag. Dr. Karl Koschutnig for his support with the MRI measurement and data analyses. He was indispensable and an incredible help for this study, giving me methodological and theoretical advice concerning arising brain- and fMRI-related questions. In addition to this, he supported me concerning data proceeding, analyses of fMRI, and interpretational questions. Thank you so much for everything!

I want to thank Univ.-Prof. Dr. Th. Hummel for showing me his department and for explaining to me how his team conducts fMRI studies.

I thank my external supervisor Emilia Iannilli, PhD who gave me constructive advice concerning methodological questions related to the MRI set-up, especially the modification of the MRI in order to be able to administer fluids to the participants. Additionally, she supported me in data preproceeding questions, analyses, and interpretation. Thank you very much for your time and effort.

I am very thankful to Univ.-Prof. Dr. Walter H. Kaye. He was the researcher who inspired me to conduct the present study and has been very supportive throughout the past few years. I am deeply thankful to him and his team for supportive discussions and their guidance regarding the MRI measurements.

I also want to thank Univ.-Prof. Dr. Brian Lask for many supportive discussions about content-related issues.

I would like to thank all my colleagues at the Institute of Social Medicine and Epidemiology, especially Univ.-Prof. Dr. Éva Rásky, Univ.-Prof. Dr. Willibald J. Stronegger, Dr. Franziska Großschädl, MSc BSc, and Univ.-Prof. MMag. Dr. Johanna Muckenhuber for discussing arising questions concerning my thesis and giving valuable input to improve the quality of my work. Thank you very much.

The project was conducted in cooperation with the Clinic of Neuroradiology and I would like to thank the clinic staff, and especially their director, Univ.-Prof. Dr. Franz Ebner, for their support. If Univ.-Prof. Dr. Franz Ebner had not endorsed my wish to conduct a taste study, I would not have been able to do this research.

The MRI study I conducted made financial support necessary. Among other things, I needed grants to buy the necessary equipment and to modify the MRI to be able to administer fluids to the participants. Therefore, I wholeheartedly want to thank *Land Steiermark* for funding this study. Without their support, I would not have been able to implement this project. I would also like to thank the doctoral school, "Sustainable health research", at the Medical University Graz for their financial support. Without their grants I would not have been able to do this research.

Additionally, I would like to thank all clinicians who helped recruiting the study's participants, especially Dr. Claudia Bieberger, Hilde Brandtner, DSA, Univ.-Prof. Dr. Marguerite Dunitz-Scheer, Dipl.-Psych. Stefanie Gruber, Univ.-Prof. Dr. Peter Scheer, and Univ.-Prof. DDr.

Michael Lehofer. I also want to thank the *BG/BRG Klusemannschule* for allowing me to introduce my study to the students there, thus enabling me to recruit healthy control women.

I am indebted to the participating women who have dedicated their time and given their effort supporting this project.

I am deeply thankful to my friends who have been there for me in the past few years, who encouraged me when I felt hopeless and supported me in stressful, but also joyful days. I would like to especially thank Nina Brand, Barbara Welten, Christiane Steinhöfler, Aurelia Berger, Daniela Raymitz, DI (FH) Ralf Wießpeiner, Mag. Clemens Schuster, Mag. Denise Nittel, Mag. Dr. Evelin Singewald, and Dr. Majda Köck-Deutsch. Additionally, I would like to thank Dr. Gertraud Diestler for always having a sympathetic ear, and Dr. Birgit Kirsten Steinbrenner for her endless support in all the good and bad times of my life. Thank you all so much.

Most of all, I dearly want to thank my parents for being there for me for my entire life in such a supportive and loving manner, emotionally, intellectually, and financially. I will always love you.

I want to thank my beloved sisters, Dr. Nicole Weirich, Mag. Renée Georgiev, and Désirée Burkert-Sridhar, LL.B. You are not only my family, but also my soul-mates. You are in my heart forever.

I would also like to thank my nephew and nieces – Julian and Marie, Kaiya and Miya, Marietta and Carlotta – for letting me see the world through their wonderful eyes.

I would like to thank my deeply loved dog Fluffy who loves me the way I am, who is always there for me and is the best friend I have ever had. You will always be in my heart.

With all my love I want to thank my beloved husband, companion and friend DI (FH) Albert who has always been there for me in good and bad times in the past few years, accepts and loves me, has been supportive and understanding and who I can laugh with – even though times might be stressful. I love you more than anything.

Last but not least, I want to thank my dear son Frederick who I love in all respects. He brought light into my life and always gives me hope, joy, happiness, and love. I love you more than anything forever.

Contents

Additional material (appendix: declaration of consent and questionnaires in German) can be downloaded at the webpage of this book at www.springer.com.

Abbreviations

ACC	Anterior cingulated cortex
AI	Anterior insula
AN	Anorexia nervosa
BA	Brodmann area
BDI	Beck Depression Inventory
BMI	Body Mass Index
BN	Bulimia nervosa
BOLD	Blood oxygenation level dependent
CA	Cornu ammonis areas
CC	Corpus callosum
CW	Healthy control women
DG	Dentate gyrus
DLPFC	Dorsolateral prefrontal cortex
DSM-IV	Diagnostic and Statistical Manual of Mental Disorders
DTI	Diffusion tensor imaging
ED	Eating disorder
EDs	Eating disorders
EDI-2	Eating disorder inventory 2
FBeK	Questionnaire concerning attitudes toward ones own body
FPI-R	Freiburger Personality Inventory
fMRI	functional Magnetic Resonance Imaging
FC	Frontal cortex
FG	Frontal gyrus
FO	Frontal operculum
FWE	Family wise error
GABA	Gamma aminobutyric acid
HZI-K	Hamburger Obsessive-Compulsive Behavior Inventory
IPL	Inferior parietal lobule
ICD-10	International Classification of Diseases
LV	Lateral ventricle
M	Mean

MANOVA	Multivariate analyses of variance
MR	Magnetic Resonance
OFC	Orbifrontal cortex
PCG	Postcentral gyrus
PET	Positron Emission Tomography
PFC	Prefrontal cortex
PHG	Parahippocampal gyrus
PTSD	Post-traumatic stress disorder
SD	Standard deviation
SES	Socioeconomic status (class)
RAN	Women recovered from anorexia nervosa
RBN	Women recovered from bulimia nervosa
ROI	Region of interest
SCL-90R	Symptom checklist
SIB	Self-injurious behavior
SITBI	Suicidal aspects and self injurious behavior
SPM	Statistical parametric mapping
STAI	State trait anxiety inventory
SVF-48	Coping questionnaire
T1R3	Type 1 member 3
WHO-QoL-bref	World Health Organization Quality of Life questionnaire

List of figures

List of tables

Zusammenfassung

Essstörungen (ED) sind in den westlichen Ländern eine der weitverbreitetsten und schwersten psychischen Erkrankungen. Eine Anorexia Nervosa (AN) zeichnet sich durch einen massiven Gewichtsverlust aus, der durch eine Restriktion des Essverhaltens herbeigeführt wird. Charakteristika der Erkrankung sind ein anormales Essverhalten, Gewichtskontrolle, Körperschemastörungen und eine beeinträchtigte Stimmung. Studien konnten Veränderungen in der Größe des Gehirns in einzelnen Gehirnarealen bei PatientInnen mit einer Essstörung belegen. Zusätzlich konnte man bei PatientInnen, die an einer AN leiden, funktionelle Gehirnveränderungen in Regionen feststellen, die für die Regulation des Essverhaltens zuständig sind. PatientInnen mit einer AN zeigten auch nach der Heilung noch eine veränderte Gehirnaktivität in der Insula, dem orbifrontalen Cortex, dem mesialen temporalen, parietalen und anterioren cingulären Cortex als Reaktion auf die Verabreichung eines Geschmacksreizes. Bei PatientInnen mit AN könnte daher die Veränderung der Gehirnaktivität bei der Antizipation von Essensreizen Verhaltensstrategien zur Vermeidung von Nahrung induzieren. Bisher wurde bei PatientInnen, die an einer Anorexie leiden, in den meisten Untersuchungen lediglich die Gehirnaktivität als Reaktion auf einen süßen Geschmacksreiz (Zuckerlösung, Kakao und Milch-Schakes) untersucht, da man annimmt, dass dies ein positiver Stimulus ist. Daher war es das Ziel der vorliegenden Untersuchung, psychologische Faktoren, die mit der Erkrankung einhergehen sowie strukturelle Gehirnveränderungen und Veränderungen in der Gehirnaktivität bei der Verarbeitung von Nahrungsreizen als Reaktion auf verschiedene Geschmacksreize (süß, sauer und herzhaft) zu untersuchen sowie potenzielle Assoziationen zwischen den neurobiologischen Veränderungen und den psychologischen Faktoren zu überprüfen.

Eine Gruppe von 21 Frauen, die an einer Anorexia Nervosa litten und 21 gesunde altersgematchte Frauen (CW) wurden untersucht. Daten hinsichtlich psychologischer Variablen wurden mittels Fragebogen erhoben. Die Größe des Gehirns und funktionelle Gehirndaten wurden mit einem 3 Tesla Magnetresonanzscanner (MRI) während der Verabreichung von drei verschiedenen Geschmacksreizen (150mMol Zuckerlösung, 50mMol Zitronensäurelösung und 50mMol Monosodium Glutamatlösung) im Vergleich zu einem neutralen Geschmacksreiz gemessen. Die Daten hinsichtlich psychologischer Aspekte der Erkrankung wurden mittels Multivariater Varianzanalyse (MANOVA) analysiert. Unterschiede in den Volumina der Gehirnareale zwischen den Gruppen AN und CW wurden mittels Varianzanalyse, Unterschiede in den hippocampalen Substrukturen mittels MANOVA untersucht. Zusätzlich wurden Korrelationen zwischen der Größe des Hippocampus mit Stress, positiven und negativen Stressverarbeitungsstrategien berechnet.

Mittels der Software SPM wurden die funktionellen Gehirndaten analysiert. Nach der Vorverarbeitung der Daten wurden Gruppenunterschiede in der Gehirnaktivität als Reaktion auf einen Geschmacksreiz im Allgemeinen untersucht. Basierend auf diesen Ergebnissen wurde in einem weiteren Schritt eine *Region of interest* (ROI)-Analyse durchgeführt. Hierzu wurden Unterschiede in der Gehirnaktivität zwischen den beiden Gruppen AN und CW als Reaktion auf einen süßen, sauren oder herzhaften Geschmacksreiz in den ROIs mittels MANOVA untersucht. Zusätzlich wurden Korrelationen zwischen der durchschnittlichen Beta-Aktivität als Reaktion auf die Verabreichung eines Geschmacksreizes im Allgemeinen in den ROIs mit psychologischen Daten wie beispielsweise Stress, der Beurteilung der Geschmäcker, Ko-Morbiditäten und der Erkrankungsdauer berechnet.

Die Ergebnisse dieser Untersuchung zeigten, dass Patientinnen mit einer AN spezielle Persönlichkeitscharakteristika wie Emotionalität, Perfektionismus oder soziale Unsicherheit aufweisen. Viele der untersuchten Patientinnen litten an den klassischen Ko-Morbiditäten Depressionen, Angst- oder Zwangsstörungen. Die Frauen in der Gruppe AN beurteilten ihren subjektiven Gesundheitszustand sowie ihre Lebensqualität niedriger als jene in der CW-Gruppe. Zusätzlich zeigten sich Defizite in der Stressverarbeitung und Störungen in der Körperwahrnehmung bei Patientinnen mit einer AN.

Bei Patientinnen mit AN wurde ein selektiver Volumensverlust in speziellen Gehirnregionen (dem kortikalen Gehirnvolumen insgesamt, in der grauen Gehirnmasse, dem Corpus callosum, dem Thalamus, dem Choroid Plexus und in der Amygdala) festgestellt. Es zeigten sich Zusammenhänge zwischen der Größe des Hippocampus mit psychologischen Variablen wie Stress und Stressverarbeitungsstrategien. Zusätzlich wurden auch Unterschiede in der Geschmacksverarbeitung in speziellen Gehirnregionen, die für Belohnungsverhalten, das Fällen von Entscheidungen, für emotionale Reaktionen, Kontrollverhalten und für die Integration verschiedener Informationen verantwortlich sind, wie beispielsweise der Insula, dem anterioren cingulären Cortex oder dem orbifrontalen Cortex, unabhängig von der Geschmacksmodalität, festgestellt. Die Untersuchung zeigte auch einen Zusammenhang zwischen der Geschmacksverarbeitung mit psychologischen Variablen wie Stress, Ängstlichkeit, Ko-Morbiditäten und der Erkrankungsdauer auf.

Aufgrund der Unzugänglichkeit des Gehirns sind bislang kaum Informationen über physiologische Korrelate von Verhaltensstörungen verfügbar. Bildgebende Verfahren ermöglichen die Gewinnung neuer Erkenntnisse über strukturelle und funktionelle Gehirnveränderungen, die – in Interaktion mit psychologischen und sozialen Faktoren sowie Umwelteinflüssen – zur Entwicklung und/oder Aufrechterhaltung von Essstörungen beitragen können. Insgesamt fördern die Ergebnisse der Untersuchung ein besseres Verständnis der Pathophysiologie der Erkrankung, was dazu beitragen kann, einen darauf begründeten Ansatz für die Therapie zu entwickeln.

Abstract

Eating disorders (EDs) are among the most widely spread and most severe mental diseases in Western countries. Individuals suffering from anorexia nervosa (AN) show an incredible loss of weight due to a restriction of food intake. The disease is associated with abnormal eating behavior and weight regulation, body image distortion and mood disturbances. Previous studies have evidenced a reduced volume in special cortical regions in individuals suffering from an ED. Additionally, functional alterations can be found in cortical regions in patients suffering from AN, which are implicated in the regulation of feeding behavior. Patients with AN – and even recovered subjects – show altered activity in the insula, the orbifrontal cortex, the mesial temporal, parietal, and the anterior cingulated cortex compared to healthy individuals when a taste stimuli is administered. Cortical enhancement of anticipatory signals related to food could therefore trigger behavioral strategies to avoid the exposure to food in AN. So far, most studies concerning patients suffering from AN analyzed brain activity in response to a sweet taste (sucrose, cocoa, milkshakes), assuming that this taste is rated as being pleasant. Therefore, the aim of this study was to analyze psychological factors which are associated with the disease, structural brain changes, possible alterations in brain activity in response to different tastes (sweet, sour, and umami), as well as correlations between neurobiological changes and psychological factors.

Twenty-one females who were currently suffering from AN, and 21 healthy age-matched control women (CW) were tested. Psychological data was measured with a questionnaire. Information regarding cortical volume was collected, together with imaging data from a 3 Tesla magnetic resonance scanner (MRI), while three different taste stimuli (150mMol sucrose-solution, 50mMol citric acid-solution, and 50mMol monosodium glutamate-solution) as well as a neutral solution (artificial saliva) were presented to the participants. Data concerning psychological aspects of the disease (i.e. personality, stress, coping, co-morbid disorders) was analyzed using multivariate analyses of variance (MANOVA) for each dimension. Brain volumes were analyzed by means of analysis of variance and the volumes of hippocampal subfields were analyzed using MANOVA. Correlations were calculated between the hippocampal volume and aspects such as stress or positive and negative coping strategies.

The brain activity in response to different taste stimuli was analyzed with the statistical software SPM. After pre-processing, group differences due to the administration of a taste in general were analyzed. Based on these results, a region of interest (ROI) analysis was performed, calculating differences in brain activity between the AN and CW groups concerning the administration of the three different tastes by means of MANOVA. In addition to this, Pearson's correlation coefficients were calculated between the mean beta activity in the ROIs and psychological data such as stress, anxiety and pleasantness ratings, co-morbidities, and the duration of the illness.

The results of this study evidenced that patients with AN show specific personality characteristics such as high emotionality, perfectionism, or social insecurity. Many patients with AN suffer from the associated co-morbid disorders (depression, anxiety disorder, or obsessive-compulsive disorder). Individuals with AN rated their subjective health worse and had a lower quality of life. Additionally, the results of this study evidenced a deficiency in coping strategies as well as body image distortions in AN.

Patients with AN showed a selective volume loss in special brain areas (overall cortex volume, gray matter volume, corpus callosum, thalamus, choroid plexus, and amygdala), and the size of the hippocampus was associated with psychological factors such as stress and deficiencies in coping. Moreover, the results of this study evidenced altered taste processing in brain areas such as the insula, the anterior cingulated cortex, and the frontal cortex, being relevant for reward, decision-making, emotional responses, control, and the integration of information, irrespectively of taste modality. In addition to this, the results showed that taste processing is associated with stress, anxiety, co-morbidities, and the duration of the illness in AN.

Up until now, hardly any information regarding physiological correlates of behavioral disorders was available because of the inaccessibility of the brain. Imaging methods can provide new insights into the structural and functional changes of the brain which – in addition to psychological, social and environmental interactions – might contribute to the development and/or maintenance of AN. Overall, the results of this study help to better understand the pathophysiology of the disease which can help to find a reasoned approach to its treatment.

1 Scientific background

1.1 Introduction

Eating disorders (EDs) are among the most widely spread and severe diseases in the Western countries and mainly affect young women (Bergmann et al., 2003; Eberly, 2005; Makowski et al., 2014). EDs and especially anorexia nervosa (AN) are frequently debilitating, with the highest mortality rate of all mental disorders (Bulik et al., 2006; Frank et al., 2008). The disorders are of unknown aetiology and begin in puberty in young women (Barbarich et al., 2003). Epidemiological studies have shown that the age of onset of AN is between 15 and 19 years of age (Watkins, 2011). Prevalence studies show that approximately 0.9-2.2 % of women in Western countries suffer from AN during their lifetime, and between 2.4 and 4.3 % suffer from partial-syndrome AN (Watkins, 2011). The disease is associated with abnormal eating behaviors, body image distortions and mood disturbances (Tejado et al., 2010). The general clinical characteristics that define AN are the persistent desire to stay extremely thin, a pathological fear of weight gain combined with a distortion of one's own body perception. These core symptoms are accompanied by specific personality characteristic (Tejado et al., 2010). Although EDs differ concerning many aspects, they also share several common problems such as the preoccupation with weight and food, life in the extremes, suppression of emotions, and hypersensitivity (Kaye et al., 2005).

The development of an ED is often attributed to the cultural environment, e.g. the effects of media on weight, shape and body image. In fact, a relationship between the perceived pressure to be thin and body image dissatisfaction can be found in Western countries (Watkins, 2011), but while all women in today's society are exposed to influences which encourage losing weight, only some develop an ED (approximately 15.4 % of female high school and college students have an ED) (Lemberg, 1999). It is thus thinkable that underlying biological mechanisms might trigger the progress of the disease (Kaye et al., 1998). Pubertal-related female gonadal steroids, age-related changes, stress and/or cultural and social pressures may contribute to the onset of an ED in puberty (Kaye et al., 2005). Twin studies showed that about 50 to 80 % of the variance in EDs can be explained by family factors (Kaye et al., 2009). Additionally, specific personality traits can lead to a predisposition to develop an ED (Kaye et al., 1998, 2005, 2009; Lemberg, 1999). However, it remains unclear whether subjects suffering from an ED have a primary disturbance that modulates feeding or whether a disturbed appetite is secondary to other phenomena such as anxiety and occupation with weight gain (Wagner et al., 2008).

According to ICD 10, EDs are classified as behavioral syndromes associated with physiological disturbances and physical factors, and one of the main criteria of anorexia as well as bulimia nervosa (BN) is the existence of a body distortion which goes along with the belief and perception of being overweight and an intense criticism or hatred of parts of the body (Ehrig, 2003; Espinoza et al., 2009; Lemberg, 1999; Robertson, 2009; Watkins, 2011).

1.2 Anorexia nervosa

Anorexia is a disorder of unknown aetiology that affects mostly young women from the onset of puberty (Barbarich et al., 2003). The illness usually starts in mid-teen years with a dietary restriction (Fairburn & Harrison, 2003), and is more common in higher socioeconomic classes (SES). The lower acceptance of obesity in families with a higher SES might be a factor explaining the higher incidence of AN in these classes (Lindberg & Hjern, 2003). The disease is connected to an abnormal eating behavior, to weight regulation, body image distortion and mood disturbances (Barbarich et al., 2003; Kaye et al., 2005; Lemberg, 1999; Watkins, 2011). It is related to severe medical complications, nutritional and endocrine alterations (Castro et al., 2004), as well as brain abnormalities (Kaye, 2009). Individuals with AN have a dysfunctional cognition concerning weight and shape, combined with an incredible fear of weight gain, and an enormous obsession with fatness even in the face of cachexia (Kaye et al., 2005; Lemberg, 1999; Uher et al., 2003).

According to DSM-IV (American Psychiatric Association, 1994), AN is characterized by a self-induced body weight of at least 15 % below the expected number, a fear of gaining weight, body image distortion, and an endocrine disorder resulting in amenorrhea. Although amenorrhea is a diagnostic criteria for AN, the absence of menstrual cycles in individuals suffering from the disease has meanwhile been judged as questionably relevant (Watkins, 2011). Patients with AN frequently show a cognitive preoccupation with food in an obsessive manner. Two subtypes are specified according to DSM-IV: a restrictive and a binge/purge type. The restrictive type does not regularly engage in binge-eating or purging behavior (self-induced vomiting or misuse of laxatives, diuretics, or enemas) during the current episode of anorexia nervosa. The binge-eating-purging type regularly engages in binge-eating or purging behavior (self-induced vomiting or the misuse of laxatives, diuretics, and enemas). The diagnostic criteria of AN according to ICD-10 (World Health Organization, 1994) and DSM IV (American Psychiatric Association, 1994) are shown in table 1.

Table 1: Diagnostic criteria of AN according to ICD-10 (World Health Organization, 1994) and DSM-IV (American Psychiatric Association, 1994)

ICD-10	DSM-IV
A disorder characterized by deliberate weight loss, induced and sustained by the patient. It occurs most commonly in adolescent girls and young women, but adolescent boys and young men may also be affected, as may children approaching puberty and older women up to the menopause. The disorder is associated with a specific psychopathology whereby a dread of fatness and flabbiness of body contour persists as an intrusive overvalued idea and the patients impose a low weight threshold on themselves. Usually there is undernutrition of varying severity with secondary endocrine and metabolic changes and disturbances of bodily functions. The symptoms include restricted dietary choice, excessive exercise, induced vomiting and purgation, and the use of appetite suppressants and diuretics.	Refusal to maintain the body weight at or above a minimally normal weight for age and height: Weight loss leading to the maintenance of a body weight < 85 % of the expected number or the failure to make the expected weight gain during the period of growth, leading to a body weight of less than 85 % of that expected. Intense fear of gaining weight or becoming fat, even if underweight. Disturbance in the way a person's body weight or shape is experienced by themselves; undue influence of body weight or shape on self-evaluation and a denial of the seriousness of the current low body weight. Amenorrhea (at least three consecutive cycles) in postmenarchal girls and women. Amenorrhea is defined as periods occurring only following hormone (e.g. oestrogen) administration.

Studies found that the binge/purge type shows more psychopathological symptoms and tends to have a worse outcome than the restrictive type (Peat et al., 2009). Patients from the restrictive type cross over to the binge/purge type, but the reverse is less common (Peat et al., 2009); parental criticism is associated with crossing over in AN (Tozzi et al., 2005). Additionally, low self-directness is associated with a higher incidence of a crossover between ED subtypes.

There is evidence that the presence of vomiting in EDs is more frequent in subjects who show a higher lifetime maximum and minimum BMI (who are more prone to be overweight), earlier menarche, lower rates of laxative abuse, and lower self-directedness (Reba et al., 2005). Overall, laxative abuse in EDs seems to be associated with a worse course of illness, a higher prevalence of borderline personality disorder including suicidality and self-harm, feelings of emptiness, and anger (Tozzi et al., 2005).

Specific personality traits which can enhance the development of an ED are intensified during adolescence by multiple factors such as gonadal steroids, development, stress, and culture. In patients with AN it seems that dieting reduces dysphoric mood while eating emphasizes it. In turn, after the onset of the disease chronic dieting and weight loss lead to neurobiological changes which enhance denial, obsession, rigidity, anxiety and depression (Kaye et al., 2009).

During the development of the disease genetic and environmental factors interact. In individuals who show a specific genetic profile (concerning disturbances of noradrenergic functioning), puberty-related changes in combination with other factors such as stress, trauma, peer-group influences, and family problems can lead to the onset of the disease. In turn, anorexia can also be maintained by socio-cultural influences, stress and by the management and advantages of the illness (Nunn et al., 2011).

1.2.1 Medical issues and complications of the disease

The medical consequences of AN include vitamin deficiencies which have an impact on cognitive functioning, electrolyte imbalances, arrythmias, bradycardia, hypotension, hypothermia, muscle weakness and osteopenia. Furthermore, anorexia includes dryness and loss of hair, abnormal heart function, growth of body hair, amenorrhoe and the risk of fertility problems (Lemberg, 1999). Among the most serious complications are the refeeding syndrome, cardiac effects, and possibly irreversible changes in the bones and in the brain (Kaplan, 2005). Although nutritional rehabilitation and weight restoration are early goals in patients with AN, refeeding a patient too rapidly entails a lot of dangers. The refeeding syndrome is characterized by fluid and electrolyte problems, cardiac, hematologic and neurologic complications as well as sudden unexpected death (Golden & Meyer, 2004). Even though some medical complications are reversible, osteopenia, structural brain changes, and, in addition to this, body image distortion as well as preoccupation with food seems to persist after recovery (Golden & Meyer, 2004).

1.2.2 Psychopharmacological treatment

Studies concerning the efficiency of a psychopharmacological treatment of AN are sparsely encouraging. Additionally, patients who suffer from AN often refuse to be treated with medication (Lahousen et al., 2003). So far, no drugs seem to really benefit the treatment of the disease. A recent study showed that a psychopharmacological treatment had no significant effect on weight gain compared to placebo in the treatment of AN (de Vos et al., 2014). Only in BN, psychopharmacological treatment can reduce ED symptoms such as binge eating, vomiting, and depressive symptoms (Lahousen et al., 2003).

1.2.3 *Recovery*

Most patients recover after a mean disease duration of six years (Tejado et al., 2010), although a high number of cases become chronic or die (Kaye et al., 2009). Low body weight and early onset of the disease are factors of poor prognosis (Tejado et al., 2010). Overall, relapse in patients suffering from an ED is a significant tribute to the generally poor prognosis of these diseases. The prevalence of recovery rates and chronic courses shows a wide variety. Rates of relapse range from 9 to 65 %, depending on the definition of remission and relapse (Carter et al., 2009). A study by Diamanti et al. (2008) showed a long-term recovery rate (9 to 70 months) of approximately 63 % after nutritional rehabilitation in an inpatient unit and a failure rate of 37 %. The rehospitalisation rate was between 22 and 25 %. In addition, they found that about a third of patients suffer chronically and are repeatedly rehospitalized. Uher et al. (2003) report that in 10 to 20 % of anorexics the disease follows a chronic course leading to disabilities and elevated mortality. Other studies report that about half of the patients recover, one third improves, and about 20 % remain chronically ill (Kaye et al., 2009; Steinhausen, 2002). Favorable prognostic factors are an early age of onset combined with a short history. Unfavorable prognostic factors include a long history of the illness, vomiting, bulimia, binging/purging and obsessive-compulsive behavior (Fairburn & Harrison, 2003).

1.3 Psychological aspects of anorexia nervosa

1.3.1 *Etiopathogenetic factors*

Concerning the aspects influencing the development of an AN, many factors have been discussed and overall it is still not fully understood why someone is willing to starve her- or himself. Several factors have been found to be important in predicting the development of an ED. These include biological components (e.g. serotonin and dopamine levels), obesity, dieting and abnormal eating behavior, traumatic life events, and social pressure (Kaye et al., 1998, 2005, 2009, 2013, 2014; Lemberg, 1999).

Historically, EDs have been seen as a sociocultural phenomenon with respect to the fact that AN appears particularly often in cultures were thinness is deemed important (Watkins, 2011); however, studies show mixed findings concerning a possible association between dieting and the pressure to be thin (Watkins, 2011). Recent findings also support the importance of biological and familial factors (Kaye et al., 1998, 2005, 2009; Thornton et al., 2011; Watkins, 2011). Some authors discuss a genetic predisposition to develop an ED (Bulik et al., 2000; Espinoza et al., 2009; Kaye et al., 2009; Strober et al., 2000; Treasure & Holland, 1995), but also childhood treatment and co-socialisation seem to have an influence (Kendler & Gardener, 1998), showing significantly elevated risks in other family members to develop an ED (Bulik et al., 2006; Kaye et al., 1998, 2005, 2009; Strober et al., 2000; Thornton et al., 2011; Watkins, 2011). Although families differ, eating-disordered families overall share some common problems such as a preoccupation with weight, food, appearance, perfectionism, life in the extremes, suppression of emotions, and hypersensitivity (Lemberg, 1999). Additionally, specific personality traits can lead to a predisposition to develop an ED (Kaye et al., 1998, 2005, 2009, 2013, 2014; Lemberg, 1999; Tejado et al., 2010; Watkins, 2011). In addition to this, perinatal influences have been discussed regarding the development of the disease (Favaro et al., 2006).

According to the biopsychosocial model of health, which was developed by Engel (1977, 1980), biological, psychological and social factors together as well as their interaction have an influence on health and on the development of diseases. In the case of anorexia, social and cultural factors, genetic influences and biological factors, trauma, and family as well as social relationships have been discussed as etiologic factors (Polivy & Herman, 2002; Wood, 2011). A biopsychosocial model of AN was described by Lucas (1981). According to Lucas (1981), a biological vulnerability, psychological predisposition, and certain sociocultural influences precede the development of the disease and lead to dieting and, subsequently, to weight loss. In individuals suffering from AN, the disease leads to further physical, psychological, and emotional changes.

1.3.2 *Personality aspects*

The core symptoms in AN are accompanied by specific personality characteristics. Moreover, specific personality traits can lead to a predisposition to develop an ED. Studies showed that characteristics such as negative emotionality, harm avoidance, perfectionism, overly compliant behavior, inhibition, reduced social spontaneity, constriction of affect and emotional expressiveness, interoceptive awareness, high self-control, social introversion, anxiety, depression, and obsessive-compulsive traits precede the onset of an ED and persist after recovery (Barbarich et al., 2003; Kaye et al., 1998, 2005, 2009, 2014; Lemberg, 1999; Tejado et al., 2010). Restrictive-type AN shows specific personality traits such as high anxiousness, high perseveration, perfectionism, intolerance, feelings of ineffectiveness, and a desire to maintain control over one's self, life and the world (Klump et al., 2000).

1.3.3 *Stress and Coping*

The development and maintenance of AN has been reported to be induced by multiple factors, e.g. genes, culture, and stress (Jappe et al., 2014). Indeed, a study by Seed et al. (2000) found a significantly enhanced cortisol secretion in AN, which indicates a higher stress level in these patients. Moreover, studies were able to show that anorexia is associated with dysfunctional coping strategies in stressful situations (Bloks et al., 2004; Brytek, 2006; Nakahara et al., 2000; Taylor & Stanton, 2007; Troop et al., 1998). Patients with AN cope with feelings through their eating behavior, i.e. sadness and fear are managed through restrictive eating and purging, anger through self-control, self-harm and exercising (Espeset et al., 2012). In addition to this, patients with AN use coping styles such as cognitive avoidance or cognitive rumination more frequently and feel guilty more often than others (Troop et al., 1998). Swanson et al. (2010) found that patients with AN frequently use negative coping styles such as avoidance as a mechanism to manage stress. Other studies reported that in AN coping styles such as positive attitude, planning and social support seem to be impaired (Villa et al., 2009) and that patients with AN usually do not cope positively with emotional distress (Brytek, 2006).

1.3.4 *Self-injurious behavior, suicide and mortality*

Eating disorders are frequently debilitating, having the highest mortality rate of all mental disorders (Bulik et al., 1999; Frank et al., 2008). The crude mortality rate for AN is about 4 %, according to a study by Crow et al. (2009). A recent study in England reported a standardized mortality rate (SMR) of 11.5 for anorexia in young adults (age 15-24), and even higher rates for older adults (SMR of 14.0 for individuals aged 25 to 44 years) (Hoang et al., 2014). In 20-year studies the mortality rates are approximately 18 %, and in 30-year follow-ups even 20 %. The mortality rates for anorexia per year are approximately 57- to 58-times higher than expected in the population (Herzog et al., 2000; Keel et al., 2003). Nevertheless, the reported mortality rates can vary considerably in various studies as patients suffering from an ED often die due to medical complications of the disease or because of suicide. Causes of death include cardiac arrhythmias, electrolyte imbalance, inanition, the refeeding syndrome, and suicide (Lemberg, 1999).

Suicide is one of the major causes of premature death in EDs (Bulik et al., 1999). Suicide attempts in EDs are as high as 26 % (Franko & Keel, 2006; Milos et al., 2004) and a history of suicide attempts is a risk factor for completed suicide later on. Other risk factors include co-morbid alcohol abuse, mood disorders, and character traits such as impulsivity, high perfectionism, low self-directness, and the binge/purge type (Bulik et al., 1999; Foulon et al., 2007; Herzog et al., 2000; Keel et al., 2003; Milos et al., 2004; Pryor et al., 1996; Youssef et al., 2004). The risk of attempted suicide is higher in restrictive type AN patients who show a tendency of self-punishment and antisocial conduct. Patients suffering from AN are more likely to die due to suicide than those suffering from BN, suggesting that AN patients develop a habituation to pain through their self-starvation and therefore use more lethal methods in their suicide attempts (Holm-Denoma et al., 2008).

Self-injurious behavior (SIB) can be defined as "any socially unacceptable behavior, involving immediate, deliberate, direct, and usually repetitive physical injury to one's own body, resulting in mild, to moderate harm, usually without suicidal intent, and not due to psychiatric organicity" (Eberly, 2005). The prevalence of SIB in EDs ranges from 25 to 45 %. Characteristics of patients who show SIB include sexual abuse during childhood, personality disorders, neuroticism, conscientiousness, and a focus on their body (Eberly, 2005).

1.4 Neurobiological aspects of anorexia nervosa

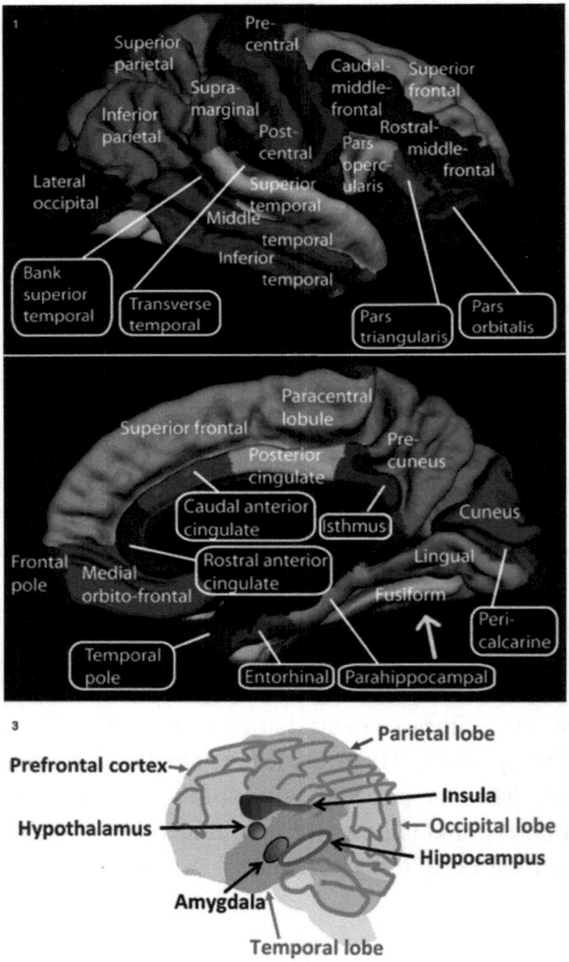

Figure 1: Brief summary of brain areas
1 Lateral surface of the cerebral cortex, 2 Medial surface of the cerebral cortex, 3 Location of the insula, amygdala and hippocampus in the brain

1.4.1 Structural changes

Alterations of brain structure and functioning were found in underweight AN (Kaye et al., 2011). MRI studies showed that patients suffering from AN have enlarged sulci and ventricles (Chui et al., 2008; Frank et al., 2004; Katzman et al., 1996; Kaye et al., 2011; Tejado et al., 2010), an enlargement of the hemispheric fissure (Artmann et al., 1985), a decreased overall brain mass (Katzman et al., 1996; Ellison et al., 1998; Frank et al., 2004; Tejado et al., 2010), and changes in the brain metabolism in frontal, cingulated, temporal, and parietal regions (Wagner et al., 2006a). Frank et al. (2013) found that patients with AN have a reduced white matter volume compared to controls. A reduction in gray matter volume in AN was found primarily in parietal and temporal areas (Castro-Fornieles et al., 2007). In addition to this, localized volume loss was found in the anterior cingulate gray matter (Muehlau et al., 2007), the pituitary gland (Kornreich et al., 1991), the cerebellum and occipito-temporal regions (Miwa et al., 2004), the middle cingulated cortex, and the precuneus (Gaudio et al., 2011). A diffusion tensor imaging (DTI) study showed a bilateral reduction of fractional anisotropy maps in the posterior thalamic radiation and the left medidorsal thalamus in patients with AN (Frieling et al., 2012). Additionally, disturbances of associational and commissural fibres seem to exist in the occipito-temporal white matter bilaterally (Frieling et al., 2012). In figure 1 a brief overview of the location of the various areas in the brain is shown.

Kazlouski et al. (2011) found deficits in the integrity of limbic and associational regions in the white matter of the brain, which indicates that body image distortion might be caused by alterations in cortical regions which are responsible for encoding body perception. Brooks et al. (2011) report that patients with AN have an enlarged dorsolateral prefrontal cortex (DLPFC) and a smaller right anterior insula (AI), as well as a reduced volume in the left and right parahippocampal gyrus, the left fusiform gyrus, the left cerebellum, and the right posterior cingulated cortex. Additionally, a reduction of total hippocampus-amygdala formation volume was found in AN (Giordano et al., 2001).

Some of the dysfunctions are inherent to deregulation of emotions or impulse control, which are also affected in these patients (Tejado et al., 2010). Positron emission tomography (PET) studies showed a global hypometabolism (Delvenne et al., 1995, 1996, 1997). Furthermore, alterations concerning neurotransmitter levels (Bailer et al., 2006; Barbarich et al., 2003; Kaye et al., 2005; Tejado et al., 2010), as well as reduced cerebral blood flow and metabolism could be observed in patients with AN (Kerem & Katzman, 2003; Tejado et al., 2010). Hypoperfusion exists in the medial prefrontal cortex and the anterior cingulated gyrus (Takano et al., 2001). Hyperperfusion was found in the thalamus and the amygdala-hippocampus complex in AN (Takano et al., 2001). In addition to this, studies were able to show that AN is associated with a significant reduction of cortical thickness (Milos et al., 2007). A ventral neurocircuit, which includes the amygdala, the insula, the ventral striatum, ventral regions of the anterior cingulated cortex (ACC), and the prefrontal cortex (PFC) is responsible for identifying the emotional aspects of stimuli and for generating an affective response (Phillips et al., 2003). A dorsal neurocircuit including the hippocampus, dorsal regions of the caudate, dorsolateral prefrontal cortex (DLPFC), parietal cortex and other regions are necessary for executive functioning, modulating attention, planning, and regulation of affective states (Phillips et al., 2003). It is possible that the changes in emotional regulation and cognition in AN are related to changes in these neurocircuits.

It is difficult, however, to differentiate between the primary disorder and the secondary phenomena accompanying the disease, in view of the fact that malnutrition originates changes in neuroendocrine functions. Whether the structural and functional abnormalities are fully reversible is still not evident. Some studies showed an increase in gray matter volume after weight restoration in patients with AN (Fuglset & Frampton, 2011; Mainz et al., 2012; Wagner et al., 2006a), others failed to find these results (Fuglset & Frampton, 2011; Giordano et al., 2001; Golden & Meyer, 2004). Studies in recovered AN showed that the white matter volume normalizes after weight restoration and the enlargement of sulci, ventricles, interhemispheric fissure, and neuropeptide function deficits seem to be reversible (Fuglset & Frampton, 2011; Kaye et al., 2011; Wagner et al., 2006a). Findings concerning a normalization of gray matter volume, however, are inconsistent (Fuglset & Frampton, 2011) and the abnormality in regional cerebral blood flow also does not seem to reverse (Frampton et al., 2011).

1.4.2 *Structural alteration of the hippocampus*

The hippocampus is a paired structure which is located in the left and right sides of the brain inside the medial temporal lobe beneath the cortical surface. It consists of different parts: the dentate gyrus, subiculum, pre- and parasubiculum, and the cornu ammonis areas (CA1, CA2, CA3, and CA4; Andersen et al., 2006). The hippocampus is part of the limbic system and plays an important role in long-term memory and spatial navigation. Anxiety, regulation of emotions, and behavioral plasticity throughout life are regulated in this area (Chowdhury et al., 2013). Many glucocorticoid receptors are located in the hippocampus and it is thus especially vulnerable to long-term stress (Joels, 2008). Stress-related steroids affect the hippocampus by reducing the excitability of hippocampal neurons, thereby inhibiting the genesis of new neurons in the dentate gyrus and causing atrophy of dendrites in pyramidal cells of the CA3 region (Andersen et al., 2006). Studies report an atrophy of the hippocampus in individuals who had suffered long-term traumatic stress (e.g. post-traumatic stress disorder (PTSD)), while other parts of their brain showed no damage or at least less damage than in the hippocampus (Andersen et al., 2006; Apfel et al., 2011; Brown & Morey, 2012; Childress et al., 2013; Dannlowski et al., 2012; Ding et al., 2013; Hall et al., 2012; Lindauer et al., 2005; Papagni et al., 2011; Rocha-Rego et al., 2012; Wang et al., 2010). Wang et al. (2010) showed PTSD to be associated with a selective volume loss in the CA3 and dentate gyrus subfields, suggesting that chronic stress suppresses neurogenesis and dendritic branching in the CA3- and dentate gyrus (DG)-structures. Like in patients suffering from PTSD, a reduction of hippocampal volume was also observed in patients with borderline personality disorder (Rodrigues et al., 2011), dissociative identity disorder (Vermetten et al., 2006), schizophrenia, and depression (Andersen et al., 2006; Campbell & Macqueen, 2004; Rodrigues et al., 2011). Connan et al. (2006) also found a significant hippocampal volume reduction in patients suffering from anorexia. Even healed AN patients seem to have a significantly reduced right and left hippocampus-amygdala formation (Giordano et al., 2001); animal experiments conducted in mice and rats support this finding by demonstrating hippocampal changes in an activity-based model of anorexia (Aoki et al., 2012; Barbarich-Marsteller et al., 2013; Chowdhury et al., 2013). CA1 pyramidal cells of mice which are vulnerable to activity-based anorexia show less Gamma-aminobutyric acid- (GABA-) -ergic innervation on cell bodies and dendrites in the hippocampus. The GABAergic innervation of hippocampal pyramidal neurons might therefore be one important cellular substrate to consider for the understanding of the progression and resilience to AN (Chowdhury et al., 2013).

Stressful life events affect the neuroplasticity of the hippocampus and associated regions. Additionally, changes concerning the neuroplasticity of the hippocampus have an influence on stress reactivity and coping (McEwen & Gianaros, 2010). Animal experiments have shown chronic unpredictable stress to induce depressive-like behaviors, passive behavioral coping and anhedonia. Stress also negatively affects the hippocampal neurogenesis in rodents (Mahar et al., 2014). Overall, the brain regions most involved in stress processing and in the regulation of associated physiological and behavioral stress processes are the hippocampus, the amygdala, and areas of the prefrontal cortex (McEwen & Gianaros, 2010).

1.4.3 *Neurobiological aspects concerning taste processing*

The brain plays a crucial role in the decision to eat, integrating multiple hormonal and neural signals which convey information about the body's nutritional status (Schwartz et al., 2000). The brain contributes to the feeding behavior including taste memory, reward, and habituation (Frank et al., 2003). A sweet taste perception is mediated to the primary cortex by tongue receptors (mainly type 1 member 3 (T1R3)) through a neural system consisting of cranial nerves (VII, IX, X), the nucleus tractus solitarii, and the thalamic ventroposterior medial nucleus – i.e. the frontal operculum (FO) and the AI (Frank et al., 2008; Kaye et al., 2009; O'Doherty et al., 2001). The AI and the associated gustatory cortex respond to the taste of food, to rewarding processes, and to decision-making and are involved in cognitive processes concerning subjective feelings (Craig, 2009; Kaye et al., 2009). The amygdala, the ventral ACC and the orbifrontal cortex (OFC) are interconnected with the insula. Afferents are directed to the ventral striatum where cortical structures involved in cognitive strategies send inputs to the dorsolateral striatum; this is where the pleasure, motivation and cognitive aspects of tastes are modulated. The integration of all these aspects leads to the decision of whether to eat or not to eat. It seems that in individuals with AN reward response is mediated by the dorsal striatum towards a strategic approach (Kaye et al., 2009).

The insula cortex is located in the center of the cerebral hemisphere and has connections with the primary and secondary somatosensory areas, anterior cingulate cortex, amygdala, prefrontal cortex, superior temporal gyrus, temporal pole, orbitofrontal cortex, frontal and parietal opercula, primary and association auditory cortices, visual association cortex, olfactory bulb, hippocampus, entorhinal cortex, and motor cortex. The insula is involved in the processing of visceral sensory, visceral motor, vestibular, attention, pain, emotion, verbal, motor information, inputs related to music and eating, and, in addition to this, to gustatory, olfactory, visual, auditory, and tactile data. This cortical region is involved in various neuropsychiatric diseases such as mood and panic disorders, post-traumatic stress disorder, obsessive-compulsive disorder, schizophrenia, and EDs. Investigations of functions and connections of the insula cortex show that sensory information including gustatory, olfactory, visual, auditory, and tactile inputs converge on the insula cortex and it seems that this is where multimodal sensory information is integrated (Nagai et al., 2007).

1.4.4 Brain activity as a response to body images

Distortion of the body image is one of the core symptoms in EDs. It is manifested by an alteration in perception, consisting of a systematic overestimation by the patient of their body size and a cognitive dissatisfaction with their body's appearance. The cerebral systems which are involved in the processing of body images and their size are the occipito-temporal cerebral cortex (visual perception of body images) and the right parietal cortex and thalamus (constitution of the map of one's own body (body scheme)). In addition to this, the parahippocampal and fusiform gyri, which are specialized in processing the visual appearance of the human body and are both implicated in the processing of emotions, are activated (Tejado et al., 2010).

The distortion of the body image which is present in patients suffering from an ED can be explained by a difference in cerebral processing concerning body images. Corresponding reduced activity can be found in the following neural areas: the insula (which is responsible for the representation of the body scheme), dorso-lateral prefrontal cortex and cerebellum (which are involved in cognitive processing), and the parahippocampal and fusiform gyri (both involved in processing emotions). Additionally, since body images are rated as aversive by patients suffering from an ED, they lead to an increase in cerebral activity at the level of the amygdala, where anxiety processing is located (Tejado et al., 2010).

In patients with AN, the processing of their own body image is characterized by an absence of activation in the before mentioned regions, which suggests a possible suppression of the cognitive, perceptual and emotional processing of their body image (Sachdev et al., 2008). In addition to this, an activation of the insula was observed in healthy controls when viewing their own body image, a factor which was not found in patients with AN (Sachdev et al., 2008). In contrast to this, a reduced activation in the middle frontal gyri, the insula, and the precuneus and occipital regions in response to their self versus other's body images was observed in AN (Sachdev et al., 2008). In another study, an activation of the right part of the amygdala was found in patients with AN when they were confronted with their own distorted body image. This suggests that negative aversive stimuli – related to anxiety – activated the right part of the amygdala in these patients (Seeger et al, 2002). Additionally, a greater left insula and right lateral prefrontal cortex activation can be observed following body estimation size and satisfaction ratings for thinner self-images in AN (Friederich et al., 2010; Mohr et al., 2010). Uher et al. (2005) analyzed the brain activity as a response to overweight images in AN. They evidenced less activation of the fusiform gyrus, which indicates that aversive stimuli exhibit diminished cerebral activity. Some of the alterations are still present after recovery and might therefore be a trait marker for AN (Tejado et al., 2010).

1.4.5 *Cognitive processing of taste-related stimuli*

Changes in the activities of brain reward systems can be measured concerning the taste, smell, thought and sight of food (Beaver et al., 2006; Delparigi et al., 2005; Gordon et al., 2000; Gottfried et al., 2002; Hinton et al., 2004; Holsen et al., 2005; Morris & Dolan, 2001; O'Doherty, 2007; Porubska et al., 2006; Simmons et al., 2005; Small et al., 2007; St Onge et al., 2005; Tataranni & Delparigi, 2003; Wang et al., 2004). These changes are found in corticolimbic reward systems and include the ventral striatum, the insula, the amygdala and the OFC (Berthoud & Morrison, 2008; Kringelbach, 2004; Volkow & Wise, 2005). The activity in the brainstem and hypothalamus changes according to the nutritional state, but the interaction with brain reward systems also influences feeding behavior (Berthoud & Morrison, 2008; Figlewicz & Benoit, 2009; Grill et al., 2007; Kringelbach, 2004; Saper et al., 2002). The rewarding properties of food, however, are disturbed by a negative energy balance in patients with an ED (Cameron et al., 2008; Stoeckel et al., 2007).

Studies found that AN patients show an increased interference with food words in Stroop tests (Channon et al., 1988; Perpina et al., 1993) and a high recognition for food words in memory tasks, irrespectively of being hungry or satiated (Morris & Dolan, 2001; Pietrowsky et al., 2002). Furthermore, they report negative valence of food stimuli (Bossert et al., 1991; Vaz et al., 1998). Adolescents with AN show higher activation in temporal and parietal areas, especially in the temporal superior gyrus during the performance of a cognitive task (Castro-Fornieles et al., 2010). This abnormal activation correlates negatively with the body mass index (BMI) and positively with depression. After recovery, a decrease of activation in these brain areas seems to persist (Castro-Fornieles et al., 2010).

Overall, neuropsychological impairment in AN concerning cognitive functions includes visuospatial abilities (Kingston et al., 1996), memory (Fowler et al., 2006; Kingston et al., 1996), set-shifting (Tchanchuria et al., 2004), and attention (Kingston et al., 1996; Lauer et al., 1999). Starvation leads to severe changes concerning impaired concentration, alertness, comprehension, and judgement, difficulties concerning decision making, increased irritability and can induce depression (Keys et al., 1950). The deficits improve after weight restoration, but it remains unclear whether the patients fully recover (Bosanac et al., 2007; Lauer et al., 1999). Poor performance on sensomotoric speed and memory tasks improve with weight gain, but difficulties in memory recognition tasks seem to persist (Hatch et al., 2010).

A family trait effect seems to exist in EDs. In a study, analyzing set-switching patients with an ED, increased perseverative errors were found. Recovered patients and unaffected sisters perform better, but still worse than healthy controls (Holliday et al., 2005; Roberts et al., 2010). Another study showed that AN patients perform worse on cognitive tasks compared to controls. The patients' siblings were intermediate with respect to neuropsychological function (Tencono et al., 2010). These results lend support to the theory that cognitive-emotive deficits represent a putative endophenotype of AN and the findings suggest a genetic basis for abnormal emotive processing in the disease.

Brain abnormalities in AN were found concerning the viewing of pictures with images of food (Naruo et al., 2000; Santel et al., 2006; Uher et al., 2004) and due to the administration of taste stimuli (Cowdrey et al., 2011; Frank et al., 2004; 2006, 2012; Hay & Sachdev, 2011; Kaye et al., 2001; 2009; Oberndorfer et al., 2013; Radeloff et al., 2014; Vocks et al., 2011; Wagner et al., 2006b; 2008). The results of these studies are reported in the following chapters.

1.4.6 *Viewing images of food*

In healthy subjects, being confronted with high-caloric food pictures is associated with an increased activity in the OFC (Beaver et al., 2006; Frank et al., 2010; Holsen et al., 2005; Porubska et al., 2006; Schienle et al., 2009; Simmons et al., 2005; Wang et al., 2004), the ventral striatum (Beaver et al., 2006), the anterior cingulate cortex (ACC; Schienle et al., 2009), the occipital lobe (Frank et al., 2010), the amygdala (LaBar et al., 2001), the insula (Frank et al., 2010; Porubska et al., 2006; Schienle et al., 2009; Wang et al., 2004), the medial prefrontal cortex, and the dorsal striatum (Killgore et al., 2004, 2006; Rothemund et al., 2007). While the OFC and insula cortex are related to food processing and to the reward system, the occipital lobe is associated with visualizing (Frank et al., 2010). The degree of positive affects while looking at high-caloric food pictures is positively correlated with the OFC activation (Killgore & Yurgelun-Todd, 2006). In addition to this, hunger/appetite ratings are positively correlated with activation in the OFC and in the insula (Karhunen et al.,

2000; Porubska et al., 2006; Wang et al., 2004). Beaver et al. (2006) showed that reward sensitivity scores significantly predict activation of the ventral striatum, the amygdala, and the OFC during the presentation of food pictures.

An interaction between homeostatic and hedonic aspects of feeding behavior exists. Fasting biases brain reward systems towards high-calorie foods and increases activation in the ventral striatum, amygdala, AI, medial and lateral OFC in healthy adults (Goldstone et al., 2009). In patients with AN being confronted with pictures of food stimuli during satiety is associated with a stronger right occipital activation and a decreased activation of the left inferior parietal cortex, as well as with a weaker activation of the right visual occipital cortex in a hungry state (Santel et al., 2006). In both AN and BN a diminished activation of the inferior parietal lobe and the occipital cortex in response to visual food stimuli could be observed (Uher et al., 2004), as well as a greater right-sided parietal and prefrontal activation compared to healthy controls (Frank et al., 2004; Santel et al., 2006). When patients with AN see images of high-caloric drinks (aversive stimulus), a specific activation is found in the left amygdala-hippocampus region, the insula, and the anterior cingulate gyrus (Ellison et al., 1998). Frank et al. (2004) found that high-caloric food leads to a greater temporal-occipital activation in AN and is accompanied by anxiety (Frank et al., 2004).

Since a higher anterior cingulate cortex and medial prefrontal cortex activity has been observed in persons with AN – both ill and recovered individuals – when confronted with food, this might be a trait marker for AN (Frank et al., 2004; Uher et al., 2003, 2004). These areas (ACC, medial prefrontal cortex) are responsible for executive functions, decision-making, and for error monitoring while the OFC is involved in coding the reward value of a stimulus (Frank et al., 2006; O'Doherty, 2004). Insula inputs to the striatum are suggested to mediate behavior including eating, especially of high-caloric food, and play an important role in the processing of information about the physiological condition of the body (Wagner et al., 2008). Patients with AN avoid this kind of food, which is consistent with the findings of abnormal responses of insula-striatal circuits that are hypothesized to mediate behavioral responses to the incentive value of food. Therefore, the question arises as to whether individuals with AN have a specific disturbance of taste modulation, or a more generalized disturbance related to the integration of interoceptive stimuli (Wagner et al., 2008). Cognitive control in AN in response to interoceptive stimuli therefore seems to alter the striatal response, thus leading to the satisfaction of long-term goals (lose weight and avoid food) rather than short-term goals (allay hunger) (Kaye et al., 2009). Cortical enhancement of anticipatory signals related to food could trigger behavioral strategies to avoid the exposure to food in AN and this abnormal response might function as a learning signal, leading to further increased avoiding behavior (Kaye et al., 2009).

1.4.7 *Administration of taste stimuli*

Five taste perceptions can be distinguished: sour, sweet, bitter, salty, and umami. Brain imaging studies have identified regions that seem to encode food reward. While anticipation of food intake leads to a greater activation in the caudal and medial OFC, rolandic operculum, dorsolateral prefrontal cortex, posterior amygdala, insula, anterior cingulate gyrus, striatum (caudate, putamen, and nucleus accumbens), midbrain, posterior cingulate, anterior hippocampus, parahippocampal gyrus, and fusiform gyrus (Small et al., 2001; Gottfried et al., 2003; LaBar et al., 2001; Pelchat et al., 2004; Uher et al., 2006), consumption of palatable foods results in a greater activation of the caudal and medial OFC, frontal operculum, rolandic operculum, amygdala, superior and mid insula, striatum, midbrain, and anterior cingulated cortex (Liu et al., 2000; O'Doherty et al., 2001, 2002; Small et al., 2001). Many regions respond to consummatory as well as anticipatory food reward, which may be due to a learning effect, where an association is strengthened through food delivery.

The primary taste cortex is located in the anterior dorsal insula and the frontal operculum (Bender et al., 2009). The anterior ventral insula has been found to respond to oral stimuli irrespectively of the modality of tastes (Rudenga et al., 2010), but the connectivity between the insula and the feeding network (hypothalamus, ventral pallidum, and striatum) seems to be greater in salty and sweet tastes than in potentially harmful solutions (quinine, or capsaicin; Rudenga et al., 2010). The OFC is activated by pleasant as well as unpleasant tastes (O'Doherty et al., 2001). Taste stimuli with higher intensity increase the activation of the middle insula for a sweet as well as salty taste (Spetter et al., 2010). In contrast to this, activation in the amygdala has only been found to increase with saltiness (Spetter et al., 2010). The activation in brain regions increases linearly with intensity, irrespectively of the pleasantness of a taste (Cerf-Ducastel & Murphy, 2004). Umami taste perception (MSG monosodium glutamate) increases food palatability through free glutamate release (Kitamura et al., 2012). An intragastric administration of glutamate in rats activates the insula cortex, nucleus tractus solitaries, amygdala, and hypothalamus (Kitamura et al., 2012). In humans, the taste of umami activates the insula cortex and peaks of activated areas in the middle insula are close to a salty taste in healthy individuals (Nakamura et al., 2011). In addition to this, the caudolateral OFC and parts of the rostral OFC are activated by the administration of umami (de Araujo et al., 2003). McCabe and Rolls (2007) were able to show that his taste activates the medial OFC and pregenual cingulated cortex and that the activation correlates with the pleasantness of the taste.

The consumption of palatable foods leads to a greater activation in the OFC and in the frontal operculum/insula, as well as to a greater release of dopamine in the dorsal striatum in healthy subjects (O'Doherty et al., 2002; Small et al., 2003; Volkow et al., 2003). Small et al. (2008) found that the anticipation of a pleasant drink results in a greater activation of the amygdala and the mediodorsal thalamus, while the receipt of the drink results in a greater activation of the left insula/operculum. Food reactivity in modality-specific sensory cortical areas seems to be modulated by internal motivational states. The left insula response to a gustatory stimulus and the brain activity in the inferior occipito-temporal region as a response to visual food stimuli tends to be stronger during fasting in healthy individuals (Uher et al., 2006). Dietary restraint scores were found to be positively correlated with an activation in the right OFC and dorsolateral prefrontal cortex bilaterally in response to a milkshake in healthy adolescents (Burger & Stice, 2011). Individuals who reported high dietary restraint had a hyper-responsivity in reward-related brain regions to food intake, which might increase the risk of over- and binge eating (Burger & Stice, 2011).

Studies have found evidence for altered functions in cortical regions in AN, which are implicated in the regulation of feeding behavior. Drinking chocolate milk when hungry activated the right amygdala and left medial temporal gyrus. In a satiated state, an activation in the inferior temporal gyrus was observed in patients with AN (and the insula in healthy controls; Vocks et al., 2011). Individuals with AN – and even recovered subjects – showed an altered activity in the insula, the OFC, the mesial temporal, parietal, and the ACC compared to healthy subjects when sucrose was administered (Kaye et al., 2009; Oberndorfer et al., 2013). A ventral (limbic) neurocircuit which includes the amygdala, the insula, the ventral striatum and ventral regions of the ACC, and the OFC, seems to be of particular relevance for the behavior of individuals suffering from AN. These regions seem to be relevant for identifying emotional aspects and the generation of affective responses to stimuli (Kaye et al., 2009). The mesial temporal cortex, which is involved in emotional processing, could lead to increased anxiety in patients who suffer from AN by an alteration of the amygdala (Frank et al., 2004). Frank et al. (2012) found that brain reward circuits in response to taste stimuli are more responsive to food stimuli in AN than in obese subjects.

Another study showed exaggerated brain responses in the ventral striatum due to the administration of a pleasant chocolate taste and in the occipital cortex due to the sight of it (Cowdrey et al., 2011). An increased response was also found in recovered AN due to the administration of an aversive strawberry taste in the insula and putamen, and in the anterior cingulated cortex and the caudate due to the sight of moldy strawberries (Cowdrey et al., 2011). In contrast to this, Radeloff et al. (2014) found a reduced response to a high-fat cream and non-caloric viscous stimulus in AN compared to BN and healthy controls.

Overall, studies found that women with anorexia show distinct differences concerning brain activity in the insula, the specific part of the brain which is important for recognizing taste (Oberndorfer et al., 2013: Wagner et al., 2008). Activity in the insula is positively correlated to odor sweetness ratings, representing the taste-like aspects of food (Veldhuizen et al., 2010). Women who have recovered from AN show significantly reduced responses in the insula and related brain regions (in large parts of the striatum and the ACC) to the administration of a taste stimulation (Wagner et al., 2008). Most recent studies, however, indicate mixed findings in regards to neurocognitve functions with recovery from AN (Hay & Sachdev, 2011). Nevertheless, evidence emerging from functional neuroimaging studies raises the possibility of a biological abnormality which may predispose to AN. There are, for example, studies which were able to show an abnormal functioning in the frontal, limbic, occipital, striatal and cerebellar regions that may persist after recovery (Hay & Sachdev, 2011). In table 2 a brief summary of studies analyzing the brain activity as a response to the administration of a taste in AN is given.

Table 2: Summary of previous studies analyzing brain activity as a response to taste stimuli in patients with AN

Study [First Author, Year, Title]	Participants	Measurements	Results
Radeloff et al. (2014) High-fat taste challenge reveals altered striatal response in women recovered from bulimia nervosa: A pilot study.	15 RAN, 14 RBN, 18 CW	high-fat stimulus, water, non-caloric stimulus	RBN showed exaggerated anterior ventral striatum activation for the cream/water in comparison to RAN or CW.

Study [First Author, Year, Title]	Participants	Measurements	Results
Oberndorfer et al. (2013) Altered insula response to sweet taste processing after recovery from anorexia and bulimia nervosa.	14 RAN, 14 RBN, 14 CW adults	sucrose and sucralose solution	Decreased right AI activation in RAN (% signal change). Main effect in the ventroposterior nuclei of the thalamus, AI, pregenual anterior cingulate, thalamus and the prefrontal gyrus.
Frank et al. (2012) Anorexia Nervosa and Obesity are associated with Opposite Brain Reward Response.	21 AN, 19 obese, 23 CW	sucrose solution vs. artificial saliva	Significantly higher activity in the striatum, the insula, the OFC and the amygdala in AN and CW compared to obese individuals.
Cowdrey et al. (2011) Increased neural processing of rewarding and aversive food stimuli in recovered anorexia nervosa.	15 RAN, 16 CW	Chocolate and moldy strawberries	RAN showed an increased neural response to the pleasant chocolate taste in the ventral striatum and to the pleasant chocolate sight in the occipital cortex. RAN had an increased activity to the aversive strawberry taste in the insula and putamen and to the aversive strawberry sight in the anterior cingulate cortex and caudate.

Study [First Author, Year, Title]	Participants	Measurements	Results
Vocks et al. (2011) Effects of gustatory stimulation on brain activity during hunger and satiety in females with restricting-type anorexia nervosa: an fMRI study.	12 AN, 12 CW	hungry vs. satiated state, chocolate milk and water	Chocolate milk in a hunger condition induced activations in the right amygdala and the left medial temporal gyrus in AN. Chocolate milk during satiety was associated with activation in the left insula in CW and in the inferior temporal gyrus on AN.
Kaye et al. (2009) New insights into symptoms and neurocircuit function of anorexia nervosa.	Review		Administering a taste to recovered AN results in a reduced blood oxygenation level dependent response in the insula, ACC and striatum. The altered activity in the supergenual ACC and medial prefrontal cortex persists after recovery.
Wagner et al. (2008) Altered insula response to taste in stimuli in individuals recovered from restrictive-type Anorexia Nervosa.	16 RAN, 16 CW	sucrose vs. neutral taste solutions	The administration of sucrose induced a reduced response in the insula bilaterally, the striatum and the ACC in RAN (% signal change). For CW, signal change was related to pleasantness ratings.

AN: women with anorexia nervosa, RAN: women recovered from anorexia nervosa, RBN: women recovered from bulimia nervosa, CW: healthy control women, AI: anterior insula, OFC: orbifrontal cortex, ACC: anterior cingulated cortex.

2 Method

2.1 Objectives

The aim of this study was to analyze psychological factors (such as personality characteristics, coping, and co-morbidities) which are present in women suffering from AN and neurobiological aspects of the disease such as structural brain changes and functional alterations concerning the processing of taste stimuli. Previous studies have evidenced specific personality characteristics which are associated with the disease (Barbarich et al., 2003; Kaye et al., 1998, 2005, 2009; Lemberg, 1999; Tejado et al., 2010). Alterations in brain volumes exist in patients with AN and it still remains unknown whether they are fully reversible or not (Frampton et al., 2011; Fuglset & Frampton, 2011; Giordano et al., 2001; Golden & Meyer, 2004; Kaye et al., 2011; Mainz et al., 2012; Wagner et al., 2006a). Additionally, previous studies have shown that an altered brain function concerning the processing of taste stimuli can be observed in subjects suffering from AN (Cowdrey et al., 2011; Frank et al., 2004; 2006, 2012; Hay & Sachdev, 2011; Kaye et al., 2001; 2009; Oberndorfer et al., 2013; Radeloff et al., 2014; Vocks et al., 2011; Wagner et al., 2006b; 2008). In most studies so far brain activity in response to taste stimuli in AN has been analyzed after the administration of sweet tastes (sucrose, cocoa, and milkshakes) (Cowdry et al., 2011; Frank et al., 2004; Hay & Sachdev, 2011; Kaye et al., 2009; Veldhuizen et al., 2010; Vocks et al., 2011; Wagner et al., 2008). Studying the brain in subjects suffering from AN after the administration of several taste stimuli and analyzing its association with various psychological aspects, i.e. pleasantness and anxiety of the stimuli or stress, can help to get a better understanding of the disease. Together with analyses of structural brain changes and psychological aspects of the disease this study can help to better understand the psychological and neurobiological mechanisms of normal as well as pathologic feeding behaviors which are present in ill subjects.

2.2 Participants

Studies have shown that in MR measurements for a liberal threshold of 0.05 about twelve subjects are required in order to achieve conclusive results (Desmond et al., 2002). As MR examinations are very sensitive to movement, it was defined that at least 18 participants of each group should be tested. The healthy control women were matched to the patients with AN in respect to their age.

Overall, 21 females suffering from AN and 21 age-matched healthy control women (CW) were tested. One woman in the AN group and one CW terminated the fMRI measurement because they got sick from the fluids which were administered; thus, the number of participants which were included into the taste analyses was 40 (20 AN and 20 CW), for all other analyses (psychological aspects and structural brain measurement) data of 21 individuals in each group was analyzed.

Patients were recruited from in- and outpatient treatment centres (i.e.. psychotherapists, in the general psychiatric ward of *Landesnervenklinik Sigmund Freud Graz,* and in the clinic of child psychosomatics at *Landeskrankenhaus Graz*) through a posting and through personal communication. The patients were recruited by advertisements specifically asking for participation of AN patients. The healthy control group was recruited by advertisements.

2.2.1 *Inclusion and exclusion criteria*

All participants had to be female, between 14 and 35 years of age and had to have a working knowledge of the German language. They were tested upon agreement to take part in the study. Written informed consent was obtained from all participants or from their parents, if they were under age. Patients suffering from AN had to be diagnosed with the illness for more than one year and their BMI in that period had to be lower than 17.5 kg/m². Their actual BMI had to be between 14.5 kg/m² and 17.5 kg/m². They had to suffer from the restrictive type of AN with no bulimic episode in their prior medical history. The Structural Clinical Interview for DSM-IV Axis I Disorders (Wittchen et al., 1997) was performed in order to assess the ED pathology. A restrictive eating pattern was defined as regularly occurring behavior such as restrictive food intake, restricting high-caloric food, counting calories and dieting. AN patients who were taking psychoactive medications other than selective serotonin-reuptake inhibitors were excluded. This criterion was selected in order to decrease heterogeneity. Those who reported binge eating or compensatory behavior in their previous history, any use of illicit drugs, and head injury with a loss of consciousness were excluded. All participants who were suffering from a psychiatric disorder (other than a co-morbid mood disorder or a higher score regarding obsessive or anxious behavior) were excluded as well.

The control women (CW) had a BMI in the normal range (between 18.5 kg/m² and 25 kg/m²) and no psychiatric illness or ED in their previous medical history. They had had normal menstrual cycles and normal weight since menarche and did not take any medication except for birth control. Additionally, the CW were not allowed to have any first-degree relative with a psychiatric disorder.

2.3 Test procedure

This study was approved by the local institutional ethical board of the Medical University of Graz (EK-number: 23-217 ex 10/11). Written informed consent from all subjects was obtained before the experiment, in case of minority it was obtained from the parents. After a clinical interview where inclusion and exclusion criteria were assessed, the study was conducted at two appointments. In the first session the participants learned how to swallow liquid while having to lie still in the 3.0 T Tim Trio system. A taste test concerning pleasantness and anxiety of the various taste solutions (sweet (150mM sucrose), sour (50mM citric acid), and savoury (50mM Monosodium Glutamate) solution) was performed. The participants rated the pleasantness and their feeling of anxiety toward the three taste stimuli. After this, their weight and height was measured and socio-demographic and medical data as well as information regarding their eating behavior and personality were collected. A comprehensive battery of clinical interviews and self-assessments was designed to assess the many aspects going along with an ED (e.g. personal, medical, behavioral, and social data). The participants filled out several questionnaires. Table 3 shows a brief description of the collected data and in table 4 the used instruments are described.

The second appointment was dedicated to the structural measurement and taste fMRI, which took approximately one hour. Before the measurement the participants were asked to rate their nervousness on a 7-point Likert scale (1 (not at all) to 7 (very much)). After the fMRI-measurement the participants rated the pleasantness and anxiety of each taste solution (150mM sucrose, 50mM citric acid, and 50mM Monosodium Glutamate solution), as well as how stressful the measurement had been for them (using a scale from 1 (not at) all to 7 (very much)). A third appointment was offered to those who wished to be informed about their results. Those participants wishing to get advice about treatment options were duly informed.

Table 3: Collected data concerning psychological aspects

Instrument	Description
Personal / sociodemographic data	sex, age, education, profession, social background and family, housing situation, social network, leisure behavior and relationships
Medical data and anamnesis	ED diagnosis, duration of the illness, age of onset, actual BMI, lowest and highest BMI, other diagnoses (e.g. depression, anxiety etc.), previous and actual treatment, medication, menstrual cycle (first menarche, regularity, anticoagulants), actual and prior eating behavior, medical complications (e.g. electrolyte disturbances, cardiovascular problems)
ED symptoms	ED symptoms – EDI-2
	body image, perceptions concerning their own body FbeK
Other psychiatric problems	depression – BDI
	anxiety disorder – STAI
	obsessive-compulsive behavior HZI
	symptom checklist SCL-90R
	suicidal aspects and self-injurious behavior (SITBI)
Personality	characteristics (FPI-R)
	stress coping questionnaire (SVF 120)
	quality of life (WHOQOL-bref)

Table 4: Brief description of all questionnaires

Test	Title, Author	Brief description
EDI-2	Eating Disorder Inventory 2, Paul & Thiel, 2005	The EDI measures the specific psychopathology of patients with an ED. The questionnaire classifies ED symptoms into the following eleven dimensions: drive for thinness, bulimia, body dissatisfaction, infectivity, perfectionism, interpersonal distrust, interceptive awareness, maturity features, asceticism, impulse regulation, and social insecurity on a 6-point Likert scale (from 0 (never) to 6 (always)).
FBeK	Questionnaire concerning Attitudes toward Ones own Body, Strauß & Richter-Appelt, 1996	The questionnaire covers the evaluation of a person's own body with 52 items. Analyses can be conducted concerning four different dimensions of body image: attractivity/ confidence, accentuation of body appearance, uncertainty/concerned, physical-sexual discomfort.
BDI	Beck Depression Inventory, Hautzinger, Bailer, Worall & Keller, 1995	The BDI measures the severity of depressive symptoms using 21 items.
STAI	State Trait Anxiety Inventory, Laux, Glanzmann, Schaffner & Spielberger, 1981	The STAI measures anxiety at the moment (state) and in general (trait), with 20 items each.

Test	Title, Author	Brief description
HZI-K	Hamburger Obsessive-Compulsive Behavior Inventory, Klepsch, Zaworka, Hand, Lünenschloß & Jauering, 1993	With this instrument, obsessive-compulsive behavior is measured concerning six different dimensions: control, washing, organizing, counting, thoughts, and to cause suffering. Additionally, a total score can be calculated reflecting the general obsessive-compulsive behavior of a person.
SCL-90R	Symptom Checklist, Franke, 1995	The SCL-90-R captures the subjective impairment due to physical and psychological symptoms in the following nine dimensions: somatization, obsession, uncertainty concerning social situations, depression, anxiety, aggressiveness, phobia, paranoid thinking, and psychoticism. All items are measured from 0 (not at all) to 4 (I suffer from the symptom very much). In addition to this, a total score can be calculated.
SITBI	Suicidal aspects and self injurious behavior (SIB), Kaess, Brunner & Plener, 2011	This questionnaire measures six different self-injurious behaviors: suicidal thoughts, suicidal plans, suicidal gestures, attempted suicide, thoughts of harming oneself, and conducting self injuries.

Test	Title, Author	Brief description
FPI-R	Freiburger Personality Inventory, Fahrenberg, Hampel & Selg, 1984	This questionnaire is an instrument which covers the following twelve dimensions of personality with 138 items: life satisfaction, social orientation, achievement orientation, inhibition, irritability, aggressiveness, strain, physical complaints, medical worries, honesty, extraversion, and emotionality.
SVF-48	Coping Questionnaire, Erdmann & Janke, 2008	The SVF-48 measures the general reaction tendency to stressful events and situations. Eight different coping styles can be recorded with the questionnaire: deemphasizing, distraction, control of situations, positive self instructions, positive use of support, escape, mental occupation, and resignation. The coping styles can be classified as positive and negative coping mechanisms.
WHO-QoL-bref	WHO Quality of Life, World Health Organization,1996	The questionnaire measures the quality of life in the following four domains: physical and psychological health, social relationships, and environment.

2.4 MRI design

2.4.1 Magnetic resonance measurement

The magnetic resonance imaging (MRI) was conducted on a 3.0 T Tim Trio system (Siemens Medical Systems, Erlangen, Germany) using a twelve-channel head coil. The participants' head was stabilized with foam cushions to prevent movements. Additionally, the participants received earplugs to limit the noise of the scanner. High-resolution anatomical images were obtained using a T1-weighted 3D MPRAGE sequence (Time of repetition TR=1900 ms, Echo Time TE=2.2 ms, 176 sagittal slices) which provides 1 x 1 x 1 mm isotropic resolution (withdrawn from Burkert et al., 2015). Functional images were obtained with a single-shot gradient echo EPI sequence (TR=2000 ms, TE=30 ms, FA=90, matrix size=64×64, slice thickness=3 mm, spatial resolution=3×3 mm). 30 transverse slices (3 mm thickness, 0.75 mm gap) were obtained parallel to a line from the anterior to the posterior commissurs (AP-PC line) in descending order. The setup for the experimental paradigm is shown in figure 2; the protocol, including a time-plan for the MR measurement, is presented in table 5.

Table 5: MR measurement protocol

Measurement	Time
Structural measurement (T1)	15 min
fMRI – Taste	30 min
Total	45 min

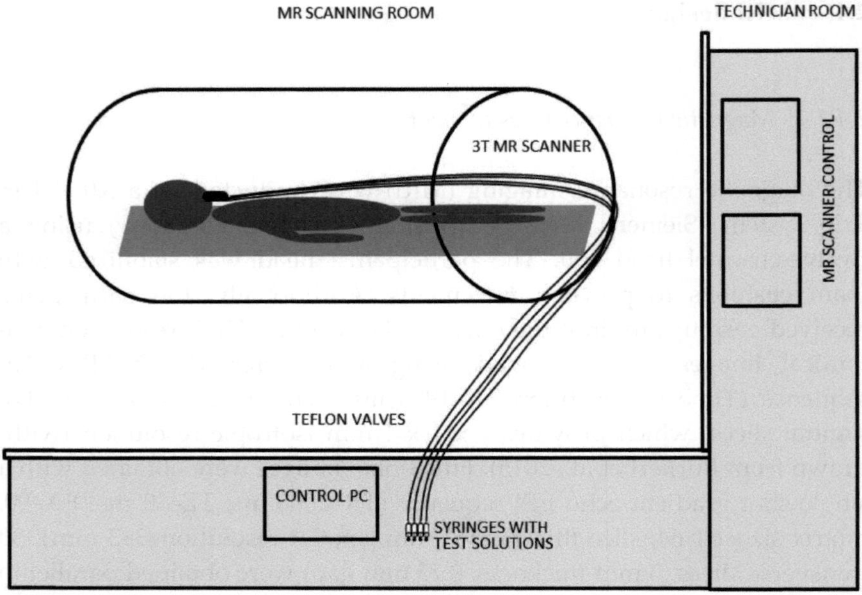

Figure 2: Experimental setup for the study

2.4.2 *Taste stimulation*

For the 24 h preceding the experimental session, the participants recorded their activity, food and fluid intake, mood, nervousness, and feelings of thirst and hunger every two waking hours. Thirst, hunger, mood and nervousness were recorded on a 7-point Likert scale with the following prompts: How hungry do you feel? (1 = not at all hungry; 7 = very hungry); How thirsty do you feel? (1 = not at all thirsty; 7 = very thirsty); How would you describe your mood right now? (1 = very good/high; 7 = very bad/low); Do you feel nervous, restless or irritable? (1 = not at all; 7 = very much so). The participants were instructed to refrain from strenuous physical activity, to abstain from alcohol for 24 hours and from caffeinated beverages on the day of the experimental session. Participants were asked to consume their regular meals on the day before the experiment, but to refrain from eating or drinking on the day of the MRI measurement for standardization purposes. All subjects were tested between 07.00 am and 10.00 am during their first ten days of the follicular phase for standardized purposes.

Three different taste stimuli compared to a neutral solution were presented to the participants: 150mMol sucrose solution, 50mMol citric acid solution, 50mMol umami solution, and a neutral one (artificial saliva: 2,5mmol $NaHCO_3$ + 25mmol KCl). To create the various solutions, sucrose, citric acid, and monosodium glutamate were dissolved in distilled water. As water activates taste cortex (Frey & Petrides, 1999; Zald & Pardo, 2000), and has a taste (Bartoshuk et al., 1964), artificial saliva was used as a tasteless rinse. The tastes were delivered to the participants in blocked presentations of 50 seconds (see figure 3). Within the run, the stimuli were presented pseudorandomysly (in the same random order for all subjects).

Every taste stimulus was delivered eight times. The solutions were presented to the front half of the tongue via Teflon tubes using purpose-built syringes. Each taste stimulus consisted of 1 ml, as very small volumes of concentrated solutions were delivered to minimize the need to swallow. Sixty ml syringes filled with the solution were connected via tubing through a wave guide to a manifold attached to the birdcage head coil in the MRI scanner. The manifold fit into the participants' mouths and delivered the taste to a consistent segment of the tongue. This procedure has been successfully used in the past to deliver liquids in the scanner (Frank et al., 2003; 2004; 2006; 2008; Kaye et al., 2009; Uher et al., 2006; Wagner et al., 2008).

The subjects received visually presented information to maintain the tastes in their mouth for 20 seconds, then they were instructed to swallow and afterwards the information "rinse" was given together with 1 ml of artificial saliva. Participants were instructed to follow the presented instructions during the entire experiment and to refrain from sucking on the tubes, swallowing and moving their head. The delivery of the stimuli was controlled and synchronized with MR scanning. After leaving the scanner, the participants rated the pleasantness and anxiety of the tastes on 7-point Likert scales (1 = dislike a lot; 7 = like a lot; 1 = very weak; 7 = very strong), and, in addition to this, they rated how stressful the situation was for them. As MRI measurement is generally a stressful situation (Mackenzie et al., 1995), this stress variable was used as an indicator for the general stress level of the participants in potentially harmful situations.

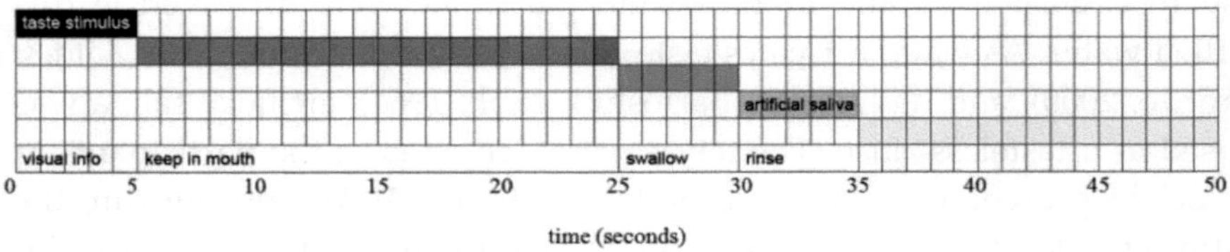

Figure 3: Set-up for taste fMRI

2.5 Data analysis

2.5.1 *Sociodemographic, medical and psychological data*

The general characteristics of the females (e.g. educational aspects, family status, age of onset of the disease, duration of illness, weight limit, health behavior, psychiatric disorders in relatives, co-morbid disorders, medical complications, treatment and medication) were analyzed descriptively for the two groups. Group differences between AN and CW concerning the age of the females, the current as well as the lowest and highest past BMI and their subjective health were analyzed by means of analyses of variance. The psychological aspects of the disease (personality characteristics, coping strategies, body perceptions, ED symptoms, depression and anxiety, obsessive-compulsive symptoms, medical complaints, and quality of life) were analyzed using multivariate analyses of variance (MANOVA) for each dimension in order to minimize multiple comparisons. Differences between the groups concerning single variables or scales were calculated by analyses of variance. Additionally, the following questions concerning self harm and suicide were analyzed by means of Chi-square tests in order to establish group differences: "Have you ever thought of committing suicide?", "Have you ever planned to commit suicide?", "Have you ever done something to make others believe you wanted to commit suicide without really wanting to kill yourself?", "Have you ever made a suicide attempt?", "Have you ever had thoughts of hurting yourself without wanting to die?", and "Have you ever hurt yourself without wanting to die?". A standard statistical software package (IBM SPSS software version 22.0 for Windows) was used. A statistical threshold of $p<0.05$ was set to reject the null hypothesis.

2.5.2 *Structural brain data*

Brain volumes were obtained using the FreeSurfer software package (Martinos Center for Biomedical Imaging, Boston, Massachusetts). An automated whole brain segmentation procedure was used to label each voxel in native space for each participant (Fischl et al., 2002). In brief, this process includes the removal of non-brain tissue, automated transformation to the Talairach space, and segmentation of subcortical gray and white matter structures (withdrawn from Burkert et al., 2015).

Between-group differences regarding the overall brain volume (total intracranial volume, white and gray matter volume, cerebrospinal fluid, supra-tentorial volume, and left and right cortex volume) as well as the volumes of 36 different brain areas (corpus callosum posterior, corpus callosum mid posterior, corpus callosum central, corpus callosum anterior, corpus callosum mid anterior, third, fourth, and fifth ventricle, left and right hippocampus, amygdala, cerebellum white matter, cerebellum cortex, thalamus, caudate, putamen, pallidum, accumbens, ventral DC, vessel choroid plexus, lateral ventricle, and inferior lateral ventricle) were analyzed using analyses of variance. A statistical threshold of $p<0.05$ was set to reject the null hypothesis.

The hippocampal subfield segmentation approach was used to calculate the volume of the following hippocampal sub-structures: presubiculum, subiculum, CA1, CA2-3, CA4/dentate gyrus (DG), fimbria, and fissure (Van Leemput et al., 2009; withdrawn from Burkert et al., 2015). This approach predicts the likelihood of the label of each voxel being assigned to a unique neuroanatomical label in native space and is based on Bayesian modeling (withdrawn from Burkert et al., 2015).

Between-group differences were analyzed regarding volumes of hippocampal sub-structures (total volume, presubiculum, subiculum, CA1, CA 2-3, CA 4/dentate gyrus (DG), fimbria and fissure) using multivariate analyses of variance in order to minimize multiple comparisons. Differences between the groups concerning sub-structures were calculated by analyses of variance (withdrawn from Burkert et al., 2015). A statistical threshold of $p<0.05$ was set to reject the null hypothesis.

Thereafter, correlations were calculated between the overall hippocampal volume and aspects such as stress or positive and negative coping strategies (withdrawn from Burkert et al., 2015). Additionally, associations between the hippocampal volume and (1) intracranial volume, (2) cortical white matter volume, (3) total gray matter volume, (4) and the duration of the illness in AN vs. CW were analyzed using Pearson's correlations. The strength of the association was interpreted as follows: correlations higher than r=0.3 and lower or equal to r=0.5 were interpreted as mid-level associations and correlations higher than r=0.5 as high-level associations (Cohen, 1992).

2.5.3 *fMRI data*

Brain activity in response to different taste stimuli was analyzed with the statistical software SPM (Statistical Parametric Mapping; Wellcome Department of Imaging Neuroscience, Lindon, UK) version 8.0 (http://www.fil.ion.ucl.ac.uk/spm/). Pre-statistical processing included motion correction, spatial normalization into the standard MNI space (Montreal Neurological Institute), and spatial smoothing with a Gaussian kernel of 8 mm FWHM (Full Width at Half Maximum). Registrations to high-resolution images were performed. The various paradigms (delivery of the four different tastes, rinsing, and fixation) were defined based on recorded time measurements during the presentation of the taste stimuli in the scanner. The statistical analyses were conducted based on the general linear model in SPM. Single-subject analyses were conducted. Afterwards, group differences due to the administration of a taste in general in a blocked design were analyzed by means of a whole brain analysis using linear t-contrasts. Clusters with a minimum of five voxel were analyzed using a FWE (family wise error)-rate-corrected statistical threshold of $p<0.05$.

Based on these results, in order to further explore the variability of the various tastes in the significant regions, a region of interest (ROI) analysis was performed. For this purpose, the mean beta activity in the significant regions for the administration of a taste in general and for the different tastes (sweet, sour and savory) was extracted and imported into SPSS software. Group differences between AN and CW in the regions of interest were calculated for the beta activity due to the administration of the different tastes using MANOVA in order to minimize multiple comparisons. A statistical threshold of $p<0.05$ was set to reject the null hypothesis.

Additionally, Pearson's correlation coefficients were calculated between the beta activity in the ROIs and (1) the rating of how stressful the fMRI-measurement was, (2) the rating of how anxious the females were concerning the administration of a taste in general, (3) the rating of how pleasant a taste was for them overall, and psychological variables such as (4) depression, (5) anxiety, (6) obsession, (7) obsessional control, and (8) the duration of illness. The strength of the association was interpreted as follows: correlations higher than r=0.3 and lower or equal to r=0.5 were interpreted as mid-level associations and correlations higher than r=0.5 as high-level associations (Cohen, 1992).

3 Results

3.1 Characteristics

3.1.1 *General characteristics*

All AN were right-handed. Eleven subjects wore either eye-glasses or contact lenses. In 38.1 % of the AN patients the highest educational level was compulsory school, 19.1 % finished an apprenticeship, 9.5 % had attended a vocational school, 19.0 % held a higher school certificate, 4.8 % finished college, and 9.5 % had a university degree. 33.3 % of participants with AN had already completed vocational training. 85.7 % of the AN patients were single, 4.8 % were in a life partnership, and 9.5 % were married. 52.4 % of AN patients still lived with their parents, 33.4 % lived alone, 4.8 % lived in a residential community and 9.5 % lived with their partner. Of all participants with AN, 23.8 % reported that they make new friends easily, 33.3 % said that they make new friends rather easily, 33.3 % reported that this is moderately difficult for them and 9.5 % said that it is rather difficult for them to make new friends. 66.7 % of the AN patients reported that they were loved by their parents.

90.5 % of the CW were right-handed and 9.5 % left-handed. Nine subjects wore either eye-glasses or contact lenses. For 38.1 % of the CW the highest educational level was compulsory school, 57.1 % had a higher school certificate and 4.8 % had a university degree. 9.5 % of the CW had already completed vocational training. 90.5 % of the CW were single and 9.5 % were in a life partnership. 38.1 % of the CW still lived with their parents, 19.1 % lived alone, 33.3 % lived in a residential community, and 9.5 % lived with their partner. Of all CW, 38.1 % reported that they make new friends easily, 52.4 % said they make new friends rather easily, 4.8 % reported that this is moderately difficult for them and 4.8 % said that it is rather difficult for them to make new friends. All CW reported that they were loved by their parents.

The patients with AN and the CW were of similar age (AN: Mean (M)=21.6 years, standard deviation (SD)=5.7 years; CW: M=21.7 years, SD=5.7 years). Overall, eight females in each group were under-age (AN: M=15.8 years, SD=1.1; CW: M=15.6 years, SD=0.6), and thirteen were between 18 and 34 years (AN: M=25.2 years, SD=4.1; CW: M=25.4 years, SD=3.7).

The AN patients had a significant lower current, lowest past, and highest past BMI compared to the CW (current BMI: AN: M=15.9kg/m^2, SD=1.4 kg/m^2; CW: M=20.2 kg/m^2, SD=1.5 kg/m^2; lowest past BMI: AN: M=13.6 kg/m^2, SD=1.7 kg/m^2; CW: M=19.3 kg/m^2, SD=1.0 kg/m^2; highest past BMI: AN: M=19.6 kg/m^2, SD=2.8 kg/m^2; CW: M=21.2kg/m^2, SD=1.9 kg/m^2). The mean age of onset of the disease in the AN group was 15.6 years and the mean duration of the disease was 5.5 years. Only four patients acquired the disease post-pubertal after menarche, while 81 % fell ill during puberty or early adulthood.

3.1.2 *Health behavior*

85.7 % of the AN patients said that they never drink alcohol, 19.0 % were smokers, and none has ever taken drugs. In contrast to this, 42.9 % of the CW said that they never drink alcohol, 38.1 % reported to drink once a week and 19.0 % drink two or three times per week. 9.5 % of the CW were smokers and none has ever taken drugs.

3.1.3 *Weight limit*

47.6 % of all participants with AN said that they do not want to gain weight over a certain limit and in 70 % of these cases the limit was lower than a BMI of 18.5 kg/m². 61.9 % of the CW said they have a personal weight limit which ranged from a BMI of 18.7 kg/m² to a BMI of 22.6 kg/m².

3.1.4 *Psychiatric disorders in relatives*

Of all the participants with AN, 9.5 % had a father with a psychiatric illness, 42.9 % had a mother with a psychiatric illness, 9.6 % had one sibling with a psychiatric illness, 4.8 % had two siblings with a psychiatric illness and 4,8 % had three siblings with a psychiatric illness. Overall, 52.4 % of the AN patients had a first-degree relative and 38.1 % had another relative suffering from a psychiatric illness.

None of the CW had a first-degree relative with a psychiatric disorder, while 38.1 % of the CW had another relative suffering from a psychiatric illness.

3.1.5 *Co-morbid disorders*

In addition to their ED, 14.3 % of AN patients suffered from an anxiety disorder, 14.3 % from an obsessive-compulsive disorder, 28.6 % from depression, 23.8 % from a sleep disorder, 9.5 % were in serious pain and 14.3 % reported to have partnership/sexual problems.

3.1.6 *Medical issues/complications*

All participants with AN suffered from at least one medical complication due to their ED: 23.8 % reported to have a low heartbeat, 52.4 % reported low blood pressure, 23.8 % suffered from constipation and 19.0 % from diarrhea, 42.9 % had circulatory complaints, 28.6 % stated to suffer from trembling, 57.1 % suffered from tiredness, 47.6 % had amenorrhea (and 14.3 % took an oral contraceptive), 52.4 % had a low body temperature, 14.3 % reported osteopenia, 28.6 % suffered from gastric or abdominal aches, 47.6 % had an electrolyte disorder, 38.1 % suffered from vertigo and 61.9 % had sleeping problems (47.6 % had problems falling asleep, 61.9 % had problems sleeping through the night).

3.1.7 *Treatment and medication*

95.2 % of the AN patients had been in an inpatient treatment program in the past, 66.7 % have had outpatient psychotherapy and 81.0 % were in a treatment program at the time of the test procedure. 33.3 % of the participants with AN took a selective serotonin-reuptake inhibitor and 14.3 % of the AN patients had taken a selective serotonin-reuptake inhibitor in the past.

3.1.8 *Subjective health*

An analysis of variance showed that a significant difference exists between the AN patients and the CW in their subjective health (p=.000). None of the participants with AN rated their health as very good, 47.6 % of them rated their health as good, 47.6 % as moderate and 4.8 % as bad. In contrast to this, 61.9 % of the CW stated their health to be very good, 28.6 % said that they were in good shape, and 9.5 % reported their health to be moderate.

3.2 Psychological aspects of anorexia nervosa

3.2.1 Personality

The results of the multivariate analyses of variance revealed significant differences in personalities between the AN and CW groups (p=.001). AN patients had significantly lower values concerning satisfaction with their life and extraversion than the CW. Additionally, they were more inhibited, more emotional and suffered from more strain and more physical complaints (table 6).

Table 6: Personality characteristics of AN vs. CW

	AN		CW		MANOVA	
Personality (FPI)	**Mean**	**SD**	**Mean**	**SD**	**F**	**p**
					4.419	**.001**
life satisfaction	5.14	2.59	7.62	2.65	9.347	**.004**
social orientation	7.76	2.32	8.81	1.57	2.935	.094
achievement orientation	7.14	2.06	5.95	1.94	3.731	.061
inhibition	7.10	3.65	4.52	2.44	7.212	**.010**
irritability	7.23	2.90	5.71	3.36	2.475	.124
aggressiveness	2.67	1.83	2.76	2.32	0.000	1.000
strain	8.10	2.74	5.10	3.18	10.750	**.002**
physical complaints	6.00	3.29	2.90	2.14	13.072	**.001**
medical worries	3.71	2.90	4.48	2.29	0.891	.351
honesty	5.57	3.04	7.24	2.30	4.010	.052
extraversion	6.10	3.69	8.24	2.83	4.468	**.041**
emotionality	7.95	3.25	5.52	3.06	6.221	**.017**

Group comparison by analyses of variance. The significant p-values are bolded. AN: women with anorexia nervosa, restrictive type (N=21), CW: healthy control women (N=21), SD: standard deviation, MANOVA: multivariate analyses of variance, FPI: Freiburger Personality Inventory.

3.2.2 *Coping Strategies*

The following results regarding coping strategies in AN vs. CW are withdrawn from a research paper by Burkert et al. (2015) and were adapted for this dissertation.

The results of the multivariate analyses of variance revealed significant differences in coping strategies between AN and CW (p=.006). Overall, AN patients use positive coping strategies less frequently than CW. Analyses of the detailed coping strategies in the positive and negative dimensions showed that AN patients less frequently use deemphasizing, situation control, positive self-instructions and less often ask for support compared to CW. No difference was found in negative coping styles between the two groups (table 7).

Table 7: Coping styles of AN vs. CW

	AN		CW		MANOVA	
Coping (SVF-48)	**Mean**	**SD**	**Mean**	**SD**	**F**	**p**
					3.362	**.006**
deemphasizing	0.87	0.60	1.69	0.65	18.142	**.000**
distraction	1.79	0.74	2.05	0.51	1.675	.203
control of situations	2.41	0.72	2.87	0.54	5.495	**.024**
positive self instructions	2.13	0.83	2.69	0.61	6.281	**.016**
positive use of support	2.47	0.80	3.09	0.70	7.149	**.011**
escape	1.85	1.04	1.55	0.88	1.026	.317
mental occupation	2.75	1.04	2.41	0.90	1.300	.261
resignation	1.71	0.99	1.37	0.76	1.571	.217
positive coping	1.80	0.47	2.33	0.34	17.147	**.000**
negative coping	2.19	0.69	2.10	0.57	0.218	.643

Group comparison by analyses of variance. The significant p-values are bolded. AN: women with anorexia nervosa, restrictive type (N=21), CW: healthy control women (N=21), SD: standard deviation, MANOVA: multivariate analyses of variance, SVF-48: Coping Questionnaire.

3.2.3 *Body perception*

The results of the multivariate analyses of variance showed that AN patients differ significantly from CW concerning body perceptions (p=.013). They have lower values concerning their perception to be attractive, are more uncertain and concerned and feel more physical-sexual discomfort (table 8).

Table 8: Body perceptions of AN vs. CW

Body perception (FBeK)	AN		CW		MANOVA	
	Mean	**SD**	**Mean**	**SD**	**F**	**p**
					3.680	**.013**
attractivity/confidence	6.90	3.21	9.71	3.05	8.455	**.006**
accentuation of body appearance	7.95	1.56	8.62	1.60	1.868	.179
uncertainty/concerned	6.57	2.56	5.10	1.92	4.466	**.041**
physical-sexual discomfort	2.86	1.35	2.05	1.02	4.785	**.035**

Group comparison by analyses of variance. The significant p-values are bolded. AN: women with anorexia nervosa, restrictive type (N=21), CW: healthy control women (N=21), SD: standard deviation, MANOVA: multivariate analyses of variance, FBeK: Questionnaire concerning Attitudes toward Ones own Body.

3.2.4 *Eating disorder symptoms*

Overall, AN patients differ significantly from CW concerning ED symptoms (p=.000). AN patients have higher scores in the scales "drive for thinness", "infectivity", "perfectionism", "interpersonal distrust", "interceptive awareness", "maturity features", "asceticism", and "social insecurity" compared to healthy controls. No difference was found concerning the domains "bulimia", "body dissatisfaction", and "impulse regulation" (table 9).

Table 9: Eating disorder symptoms in AN vs. CW

Eating Disorder Symptoms (EDI-2)	**AN**		**CW**		**MANOVA**	
	Mean	**SD**	**Mean**	**SD**	**F**	**p**
					5.558	**.000**
drive for thinness	25.86	10.54	16.86	6.11	6.698	**.013**
bulimia	11.10	4.69	11.24	3.60	.384	.539
body dissatisfaction	35.24	10.08	29.10	12.05	3.210	.081
infectivity	31.29	8.35	20.48	8.21	17.901	**.000**
perfectionism	21.43	5.78	15.81	5.60	10.245	**.003**
interpersonal distrust	21.05	6.08	13.43	4.11	22.657	**.000**
interceptive awareness	29.38	7.62	20.43	7.79	14.166	**.001**
maturity features	27.19	7.24	22.33	4.65	6.695	**.013**
asceticism	21.62	5.58	14.10	4.69	15.383	**.000**
impulse regulation	23.90	6.51	19.76	8.14	3.319	.076
social insecurity	26.00	5.36	17.14	5.68	27.039	**.000**

Group comparison by MANOVA, df=20. The significant p-values are bolded. AN: women with anorexia nervosa, restrictive type (N=21), CW: healthy control women (N=21), SD: standard deviation, MANOVA: multivariate analyses of variance, EDI: Eating disorder inventory.

3.2.5 *Depression and anxiety*

AN patients are significantly more depressive than CW, according to the Beck depression inventory. Moreover, AN patients have a higher state as well as trait anxiety than CW (table 10).

Table 10: Depression and anxiety in AN vs. CW

	AN		CW		MANOVA	
Variable	**Mean**	**SD**	**Mean**	**SD**	**F**	**p**
					12.607	**.000**
BDI	17.86	9.25	4.10	4.37	38.016	**.000**
STAI-S	51.05	12.87	33.76	8.83	25.754	**.000**
STAI-T	49.14	11.00	33.52	9.80	23.599	**.000**

Group comparison by MANOVA, df=20. The significant p-values are bolded. AN: women with anorexia nervosa, restrictive type (N=21), CW: healthy control women (N=21), SD: standard deviation, MANOVA: multivariate analyses of variance, BDI: Beck depression inventory, STAI-S: State-Trait-Anxiety-Inventory – State Anxiety; STAI-T: State-Trait-Anxiety-Inventory – Trait Anxiety.

3.2.6 Obsessive-compulsive symptoms

Results of the multivariate analyses of variance revealed no significant difference in obsessive-compulsive symptoms between AN and CW (p=.193). Nevertheless, AN patients have a higher total score in the Hamburger Obsessive-Compulsive Behavior Inventory and a higher value in the domain "organizing" than CW (table 11).

Table 11: Obsessive-compulsive symptoms in AN vs. CW

Obsessive-compulsive symptoms (HZI)	**AN**		**CW**		**MANOVA**	
	Mean	**SD**	**Mean**	**SD**	**F**	**p**
					1.544	.193
control	5.24	2.70	4.48	2.38	0.941	.338
washing	2.43	1.25	2.00	1.45	1.055	.311
organizing	4.52	2.84	2.48	2.54	6.062	**.018**
counting	2.52	2.25	1.43	1.94	2.856	.099
thoughts	3.48	2.52	2.33	2.08	2.565	.117
cause suffering	1.33	2.83	0.52	1.25	1.434	.238
HZI total score	19.52	9.79	13.24	7.46	5.482	**.024**

Group comparison by MANOVA, df=20. The significant p-values are bolded. AN: women with anorexia nervosa, restrictive type (N=21), CW: healthy control women (N=21), SD: standard deviation, MANOVA: multivariate analyses of variance, HZI: Hamburger Obsessive-Compulsive Behavior Inventory.

3.2.7 *Medical complaints*

AN patients suffer from significantly more physical and psychological symptoms than CW (p=.013). The analyses of variance showed that the participants with AN have a higher score in the domains "somatisation", "uncertainty in social situations", "depression", "phobia", "paranoid thinking", "psychoticism" as well as a higher "total score" compared to CW. No difference between AN and CW was found in the dimensions "obsession", "anxiety", and "aggressiveness" (table 12).

Table 12: Medical complaints in AN vs. CW

Medical complaints (SCL-90R)	**AN Mean**	**SD**	**CW Mean**	**SD**	**MANOVA F**	**p**
					2.838	**.013**
somatization	0.81	0.58	0.31	0.20	13.705	**.001**
obsession	0.97	0.73	0.57	0.59	3.785	.059
uncertainty in social situations	1.19	0.87	0.46	0.57	10.514	**.002**
depression	1.15	0.74	0.47	0.53	11.648	**.001**
anxiety	0.90	0.82	0.29	0.34	2.009	.164
aggressiveness	0.54	0.51	0.32	0.51	2.009	.164
phobia	0.46	0.66	0.12	0.26	4.777	**.035**
paranoid thinking	0.82	0.84	0.34	0.41	5.453	**.025**
psychoticism	0.55	0.56	0.19	0.36	6.346	**.016**
SCT total score	1.78	0.43	1.35	0.35	12.466	**.001**

Group comparison by MANOVA, df=20. The significant p-values are bolded. AN: women with anorexia nervosa, restrictive type (N=21), CW: healthy control women (N=21), SD: standard deviation, MANOVA: multivariate analyses of variance, SCL-90R: symptom checklist.

3.2.8 *Self harm and suicidality*

The results of the Chi-square tests showed that AN patients did more often plan and try to commit suicide than CW. Additionally, AN patients more often self-harmed themselves than CW. No difference was found in thoughts of committing suicide, doing something to make others believe that they wanted to commit suicide and thinking of self-harming (table 13).

Table 13: Self-harm and suicidality in AN vs. CW

SITBI	**AN**	**CW**	**X^2-Test**	**p**
Have you ever thought of committing suicide?	38.1%	19.0%	1.867	.172
Have you ever planned to commit suicide?	28.6%	0.0%	7.000	**.008**
Have you ever done something to make others believe you wanted to commit suicide without wanting to kill yourself?	9.5%	0.0%	2.100	.147
Have you ever made a suicide attempt?	19.0%	0.0%	4.421	**.035**
Have you ever had thoughts of hurting yourself without wanting to die?	33.3%	9.5%	3.535	.060
Have you ever hurt yourself without wanting to die?	28.6%	0.0%	7.000	**.008**

The percentage of subjects who answered the questions with yes is quoted. Group comparison by means of Chi-square-tests (X^2), df=20. The significant p-values are bolded. AN: women with anorexia nervosa, restrictive type (N=21), CW: healthy control women (N=21), SITBI: suicidal aspects and self-harmbehavior.

3.2.9 *Quality of life*

Overall, the AN patients report to have a lower quality of life than the CW (p=.000). AN patients have lower values in the domains "physical" and "psychological health" as well as "environment". They do not differ regarding the quality of life in the domain "social relationships" from CW, however (table 14).

Table 14: Quality of life in AN vs. CW

Quality of life (WHO-QoL)	**AN**		**CW**		**MANOVA**	
	Mean	**SD**	**Mean**	**SD**	**F**	**p**
					7.796	**.000**
physical health	14.34	1.93	17.12	2.03	20.610	**.000**
psychological health	11.68	2.86	15.46	2.24	22.687	**.000**
social relationships	15.05	2.99	16.25	2.13	2.269	.140
environment	14.69	2.03	17.06	2.26	12.762	**.001**

Group comparison by MANOVA, df=20. The significant p-values are bolded. AN: women with anorexia nervosa, restrictive type (N=21), CW: healthy control women (N=21), SD: standard deviation, MANOVA: multivariate analyses of variance, WHO-Qol: World Health Organization – Quality of Life Questionnaire.

3.3 Neurobiological aspects of anorexia nervosa

3.3.1 Structural changes in overall brain volume and volume of the various brain areas

The two groups did not differ in their total intracranial volume (p=.173) or cortical white matter volume (p=.712). Analysis of variance showed that AN patients have significantly less gray matter volume (p=.004) compared to CW. Additionally, the two groups differed in their supra-tentorial volume (p=.014), overall cortex volume bilaterally (right: p=.001, left: p=.001), their mid posterior (p=.032) and posterior (p=.019) volume of the corpus callosum, the volume of their left thalamus (p=.011), left choroid plexus (p=.014), and the volume of their amygdala bilaterally (right: p=.005, left: p=.006). In all other brain areas no differences between the two groups were found in the areas' volumes (table 15).

Table 15: Volume of the various brain areas in AN vs. CW

	AN	CW	Test statistics	
Brain volume	**Mean**	**Mean**	**F**	**p**
total intracranial	1408220	1479815	2.910	.096
brain stem	20020.8	20796.0	1.529	.223
cerebro-spinal-fluid	1187.8	1219.8	.162	.689
left cortical white matter	227200	234797	.651	.425
right cortical white matter	228228	236534	.804	.375
subcortical gray matter	164234	134549	2.329	.135
gray matter	646937	709429	11.138	**.002**
supra tentorial	999936	1076336	6.657	**.014**
left cortex	230618	260661	11.788	**.001**
right cortex	230825	262766	12.457	**.001**

Brain volume	AN Mean	CW Mean	Test statistics F	p
corpus callosum posterior	826.6	921.9	6.018	**.019**
corpus callosum mid posterior	446.9	501.7	4.965	**.032**
corpus callosum central	497.3	549.0	1.083	.304
corpus callosum anterior	824.7	875.5	1.589	.215
corpus callosum mid anterior	483.1	496.1	.075	.785
third ventricle	880.6	801.3	1.212	.278
fourth ventricle	1895.9	2125.4	2.266	.140
fifth ventricle	11.2	8.6	.636	.430
left hippocampus	4234.4	4175.2	.235	.631
right hippocampus	4209.8	4387.5	1.610	.212
left amygdala	1397.8	1579.9	8.474	**.006**
right amygdala	1433.7	1580.7	8.725	**.005**
left cerebellum white matter	14409.7	14369.1	.005	.944
right cerebellum white matter	15018.1	14853.7	.077	.783
left cerebellum cortex	57318.1	58413.5	.789	.380
right cerebellum cortex	58236.1	59939.0	1.642	.207
left thalamus	7469.7	8161.7	7.130	**.011**
right thalamus	7139.7	7443.0	2.804	.102
left caudate	3483.3	3634.9	.895	.350
right caudate	3295.1	3455.8	1.044	.313
left putamen	4895.0	4989.9	.237	.629
right putamen	4899.8	4956.5	.128	.722
left pallidum	1570.0	1546.5	.084	.773
right pallidum	1465.9	1487.1	.143	.707

Brain volume	AN Mean	CW Mean	Test statistics F	p
left accumbens	440.1	441.2	.001	.975
right accumbens	505.3	526.2	.488	.489
left ventral DC	3504.0	3546.4	.139	.712
right ventral DC	3525.9	3489.3	.130	.720
left vessel	35.0	32.5	.065	.799
right vessel	33.4	39.4	.453	.505
left choroid plexus	1149.7	928.1	6.645	**.014**
right choroid plexus	1038.6	877.1	3.366	.074
left lateral ventricle	5226.3	5546.3	.109	.743
right lateral ventricle	4899.1	5327.8	.259	.614
left inferior lateral ventricle	252.1	174.6	3.675	.062
right inferior lateral ventricle	212.7	146.3	2.482	.123

Group comparison by analyses of variance, df=20. The significant p-values are bolded. AN: women with anorexia nervosa, restrictive type (N=21), CW: healthy control women (N=21), SD: standard deviation.

3.3.2 *Volume of the hippocampal sub-structures*

The following results regarding the volume of the hippocampus and its associations with other brain areas, stress, and coping are withdrawn from a research paper by Burkert et al. (2015) and were adapted for this dissertation.

The analyses revealed no significant difference in the total hippocampal volume between AN patients and CW (p=.296). Results of the analyses of the hippocampal sub-structures showed that AN patients have a significantly smaller presubiculum, a tinier fimbria, and larger fissure compared to CW. No differences were found between the groups in the CA regions or the subiculum (table 16, figure 4).

Table 16: Differences in hippocampal volume between AN and CW

	AN		**CW**		**MANOVA**	
Hippocampal volumes	**Mean**	**SD**	**Mean**	**SD**	**F**	**p**
					4.163	**.001**
presubiculum	846.9	76.7	895.6	59.2	5.298	**.027**
subiculum	1226.7	102.8	1263.1	82.0	1.613	.211
CA1	626.3	58.9	652.5	70.3	1.722	.197
CA2-3	1869.8	246.0	1922.4	170.9	.648	.426
CA4/DG	1069.7	131.0	1075.3	92.8	.026	.874
fimbria	112.6	29.1	138.1	38.3	5.892	**.020**
fissure	96.0	36.7	65.6	17.0	11.862	**.001**

Group comparison by analyses of variance, df=20. The significant p-values are bolded. AN: women with anorexia nervosa, restrictive type (N=21), CW: healthy control women (N=21), M: mean, SD: standard deviation, MANOVA: multivariate analyses of variance.

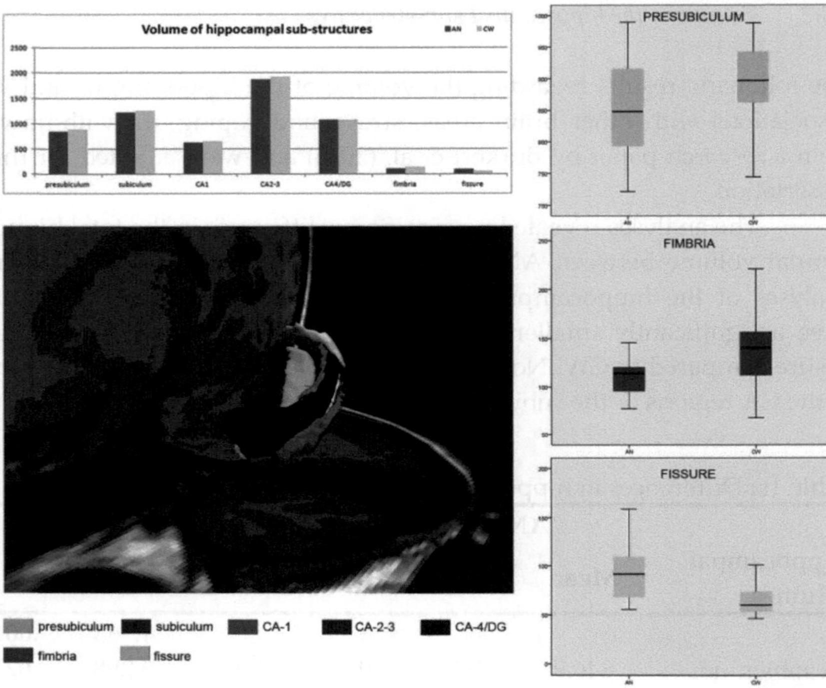

Figure 4: Volume of hippocampal subfields in AN vs. CW

The mean volumes of the seven hippocampal sub-structures are shown at the top of figure 4 in cubic millimeters. The boxplot visualizing the median and the 25 %- and 75 %-quartiles of the significant regions (the hippocampal presubiculum, fimbria and fissure) are shown on the right. AN: women with Anorexia Nervosa (N=21); NC: women in the normal control group (N=21).

3.3.3 *Associations between the total hippocampal volume and other brain areas, stress, and coping*

No association was found between the total hippocampal volume and the intracranial volume in AN patients (r=.29), but a high-level association was found between the total hippocampal volume and the cortical white matter volume in AN patients (r=.51), as well as between the total hippocampal volume and the total gray matter volume in patients with AN (r=.53). A smaller total hippocampal volume is associated with a smaller white and gray matter volume in patients suffering from AN.

In CW, a mid-level association between the total hippocampal volume and the intracranial volume was observed (r=.37). In this group no association was found between the total hippocampal volume and the cortical white matter volume (r=.14). Among the healthy control women a high level association was found between the total hippocampal volume and the total gray matter volume (r=.36). Again, a smaller hippocampal volume was related to a smaller overall gray matter volume. No association exists between the duration of the illness and the total hippocampal volume in AN (r=.10).

The analyses revealed a mid-level association between the total hippocampal volume and stress in AN patients. Smaller volumes of the hippocampus relate to more stress in patients with AN (r=-.40, figure 5). Additionally, a mid-level association was found between the hippocampal volume and positive coping in AN patients (r=-.31, figure 5), showing that smaller volumes are related to a higher use of positive strategies. The analyses revealed no associations between the total hippocampal volume and negative coping strategies in patients with AN (r=-.08).

No associations were found between the total hippocampal volume and stress or positive coping in CW (stress: r=.10; positive coping: r=.02; negative coping). A mid-level association was found between the total hippocampal volume and negative coping in CW (r=.30). Greater volumes were related to a higher use of negative coping styles.

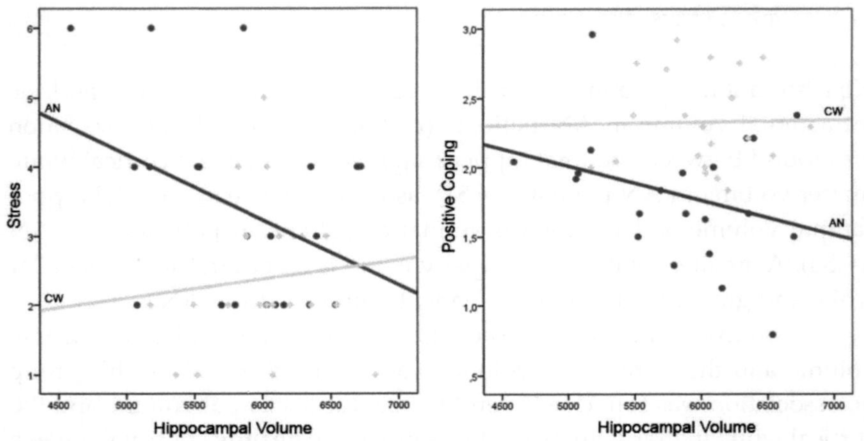

Figure 5: Association between hippocampal volume, stress and positive coping

Volume (in mm³) of the hippocampus and its associations with stress (using a scale from 1 (not at all) to 7 (very much)) and positive coping strategies (using a scale from 0 (not at all) to 4 (very often)) in AN vs. CW; AN: women with anorexia nervosa, restrictive type (N=21), CW: healthy control women (N=21).

3.3.4 *Brain activity in response to taste stimuli: whole brain analyses*

The results of the whole brain analyses in SPM showed a significant difference in the beta activity due to the administration of a taste stimuli in 24 different brain regions, including the insula, the anterior cingulate cortex, the frontal cortex, the inferior parietal lobe, Brodmann areas 1, 6, 7, 9, 34, corpus callosum, and parahippocampal gyrus. Overall, the beta activity due to the administration of a taste was higher in the AN patients than in the CW. The significant regions, the mean beta activity, the standard deviation of the beta activity and the size of the affected pixels are shown in table 17. The magnitude of the beta activity in particular regions, especially in the insula [45,-4,1], the anterior cingulate cortex [9,-46,4] and the frontal cortex [-24,32,-11] is exemplarily shown in figure 6.

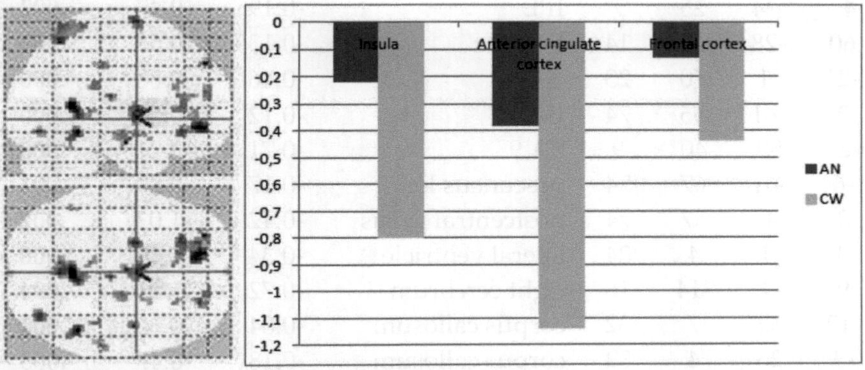

Figure 6: Beta activity due to the administration of a taste in general in the insula [45,-4,1], the anterior cingulate cortex [9,-46,4] and the frontal cortex [-24,32,-11] for AN and CW

AN: women with anorexia nervosa, restrictive type (N=20), CW: healthy control women (N=20).

Table 17: Differences between AN patients and CW in brain activity due to the administration of a taste in general: whole brain analyses

x	y	z	pi-xel	Area	AN Mean	CW Mean	p
45	-4	1	19	insula	-0.22	-0.80	**.000**
-12	29	16	32	ACC	-0.11	-0.35	**.002**
3	-52	7	71	ACC	-0.42	-1.23	**.000**
9	-46	4	71	ACC	-0.38	-1.14	**.000**
0	-10	-20	16	ACC	-0.61	0.69	**.000**
-24	32	-11	100	frontal cortex	-0.13	-0.44	**.000**
-12	47	-8	100	FG	-0.01	-0.47	**.000**
-21	47	-8	100	middle FG	-0.01	-0.28	**.000**
27	47	-8	13	middle FG	0.15	-0.14	**.000**
18	50	19	15	superior FG	-0.07	-0.39	**.002**
42	-34	25	7	IPL	-0.19	-0.47	**.002**
60	-28	40	14	BA 1	-0.13	-0.53	**.003**
-21	-1	70	23	BA 6	-0.46	-0.88	**.000**
3	-61	55	74	BA 7	-0.12	-0.51	**.000**
-27	50	40	9	BA 9	-0.71	-0.21	**.000**
-6	-61	67	74	precuneus left	-0.17	-0.89	**.001**
3	-43	67	74	postcentral gyrus	-0.42	-1.03	**.002**
3	-1	4	34	lateral ventricle	-0.34	-1.12	**.000**
9	-1	-14	16	right cerebrum	-0.72	0.06	**.004**
-12	32	7	32	corpus callosum	-0.04	-0.30	**.001**
-3	26	4	32	corpus callosum	-0.18	-0.47	**.003**
-6	-37	4	6	corpus callosum	-0.30	-0.87	**.004**
21	-13	-14	23	PHG	-0.19	-1.07	**.004**
24	5	-14	5	BA 34	-0.25	-0.81	**.020**

Significant regions concerning differences in the brain activity between AN patients and CW due to the administration of a taste; FWE-corrected with a minimum of five voxel. The significant p-values are bolded. AN: women with anorexia nervosa, restrictive type (N=20), CW: healthy control women (N=20), SD: standard deviation, ACC: anterior cingulate cortex, FG: frontal gyrus, IPL: inferior parietal lobule, BA: Brodmann area, PHG: parahippocampal gyrus.

3.3.5 *Brain activity in response to the different taste stimuli in the regions of interest*

The results of the multivariate analyses of variance showed no significant main effect due to the administration of neither sucrose (p=.116), umami (p=.253), nor citric acid (p=.087). However, results of the univariate analyses illustrated that in the insula [45,-4-1], the anterior cingulate cortex [9, -46,4] [0,-10,-20] the frontal cortex [-24,32,-11], and in the corpus callosum [-12,32-7] [-3,26,4] the AN patients showed a higher activity due to the administration of either sucrose, umami, or citric acid than the CW. Additionally, in the middle [27,47,-8], superior [18,50,19] and frontal gyrus [-12,47,-8] AN the patients had a higher activity due to the administration of sucrose and umami than the CW. In one region of the anterior cingulate cortex [3,-52,7], in the inferior parietal lobe [42,-34,25], the lateral ventricle [3,-1,4] and in the parahippocampal gyrus [21,-13,-14] the AN patients had a higher activity as a response to the administration of umami and citric acid than the CW. All results are shown in table 18. In tables 23 to 25 in the appendix the mean beta activity, the standard deviation of the beta activity and the test statistics are shown for the regions of interest for sucrose, umami, and citric acid. Figure 7 shows the magnitude of the beta activity in particular regions, especially in the insula [45,-4,1], the anterior cingulate cortex [9,-46,4] and the frontal cortex [-24,32,-11] for the three different tastes.

Table 18: Test statistics concerning differences in the brain activity due to the administration of sucrose, umami, or citric acid between AN patients and CW in the regions of interest

x	y	z	Area	Sucrose	Umami	Citric acid
				p	p	p
45	-4	1	insula	**.030**	**.014**	**.008**
-12	29	16	ACC	.052	**.047**	.129
3	-52	7	ACC	.080	**.023**	**.034**
9	-46	4	ACC	**.048**	**.014**	**.037**
0	-10	-20	ACC	**.008**	**.014**	**.002**
-24	32	-11	frontal cortex	**.020**	**.019**	**.033**
-12	47	-8	FG	**.022**	**.006**	.094
-21	47	-8	middle FG	.094	.051	.518
27	47	-8	middle FG	**.027**	**.048**	.386
18	50	19	superior FG	**.025**	**.033**	.127
42	-34	25	IPL	.086	**.041**	**.037**
60	-28	40	BA 1	.110	.115	.057
-21	-1	70	BA 6	.118	.072	.139
3	-61	55	BA 7	.097	.094	.096
-27	50	40	BA 9	**.014**	.061	.052
-6	-61	67	precuneus left	.076	.083	.055
3	-43	67	postcentral gyrus	.073	.117	.076
3	-1	4	lateral ventricle	.052	**.041**	**.020**
9	-1	-14	right cerebrum	**.047**	.097	**.020**
-12	32	7	corpus callosum	**.017**	**.009**	**.040**
-3	26	4	corpus callosum	**.047**	**.027**	**.037**
-6	-37	4	corpus callosum	.212	.067	.163
21	-13	-14	PHG	.068	**.043**	**.043**
24	5	-14	BA 34	.097	.102	**.048**

Results of MANOVA. The significant p-values are bolded. AN: women with anorexia nervosa, restrictive type (N=20), CW: healthy control women (N=20), ACC: anterior cingulate cortex, FG: frontal gyrus, IPL: inferior parietal lobule, BA: Brodmann area, PHG: parahippocampal gyrus.

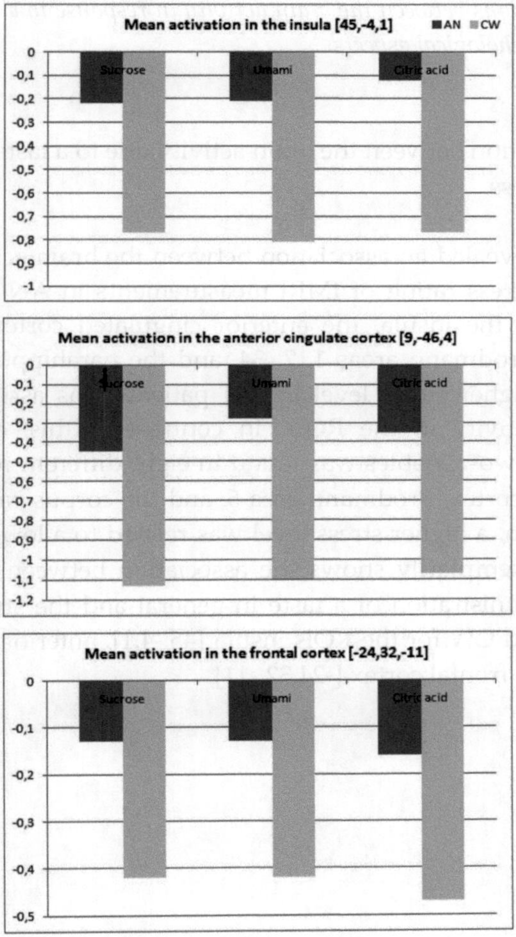

Figure 7: Beta-activity due to the administration of sucrose, umami, or citric acid in the ROIs insula [45,-4,1], anterior cingulate cortex [9,-46,4] and frontal cortex [-24,32,-11] for AN and CW

AN: women with anorexia nervosa, restrictive type (N=20), CW: healthy control women (N=20).

3.3.6 *Associations between the brain activity in response to a taste in general and psychological aspects*

3.3.6.1 Association between the brain activity due to a taste in general and stress

Our analyses revealed an association between the brain activity due to a taste and the stress rating of fMRI measurements in AN patients in 14 ROIs including the insula, the anterior cingulated cortex, the inferior parietal lobe, Brodmann areas 1, 7, 34, and the parahippocampal gyrus (table 19). A higher stress level in AN patients was associated with a lower brain activity in the ROIs. In contrast to this, an association between these two variables was found in eight different regions including the frontal cortex, Brodmann area 6, and the corpus callosum in CW (table 19). Again, a higher stress level was related to a lower brain activity. Figure 8 exemplarily shows the association between brain activity due to the administration of a taste in general and the stress rating for AN patients and CW for the ROIs insula [45,-4,1], anterior cingulate cortex [9,-46,4] and frontal cortex [-24,32,-11].

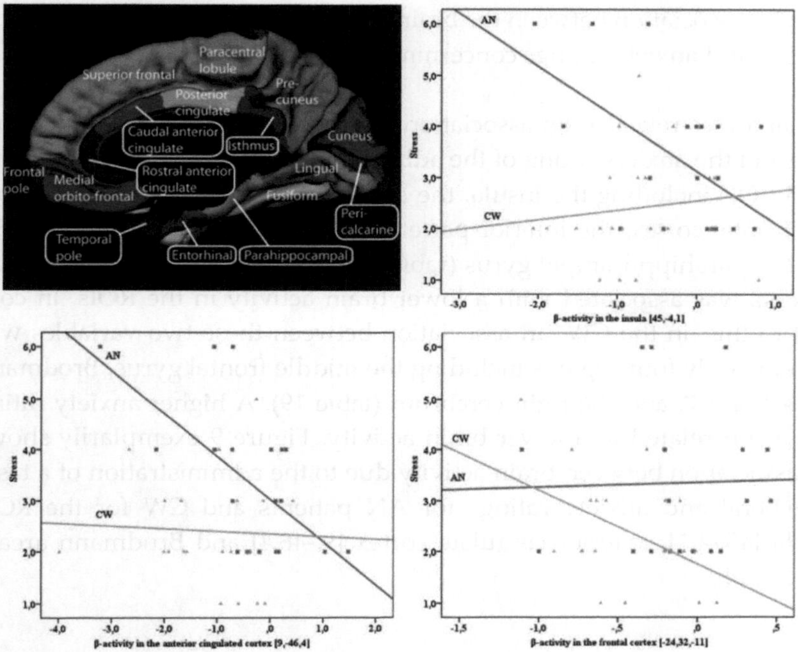

Figure 8: Association between brain activity due to the administration of a taste in general and stress in the ROIs insula [45,-4,1], anterior cingulate cortex [9,-46,4] and frontal cortex [-24,32,-11] for AN and CW

Beta-activity in the brain and its associations with stress (using a scale from 1 (not at all) to 7 (very much)) in AN vs. CW; AN: women with anorexia nervosa, restrictive type (N=20), CW: healthy control women (N=20).

3.3.6.2 Association between the brain activity due to a taste in general and anxiety ratings concerning the taste

Our analyses revealed an association between the brain activity due to a taste and the anxiety rating of the administration of a taste in AN patients in 15 ROIs including the insula, the anterior cingulated cortex, the superior frontal cortex, the inferior parietal lobe, Brodmann areas 1, 7, 9, 34, and the parahippocampal gyrus (table 19). A higher anxiety rating in AN patients was associated with a lower brain activity in the ROIs. In contrast to this, in the CW, an association between these two variables was found in only four regions including the middle frontal gyrus, Brodmann areas 1 and 7, and the right cerebrum (table 19). A higher anxiety rating was again related to a lower brain activity. Figure 9 exemplarily shows the association between brain activity due to the administration of a taste in general and anxiety ratings for AN patients and CW for the ROIs insula [45,-4,1], anterior cingulate cortex [9,-46,4] and Brodmann area 1 [60,-28,40].

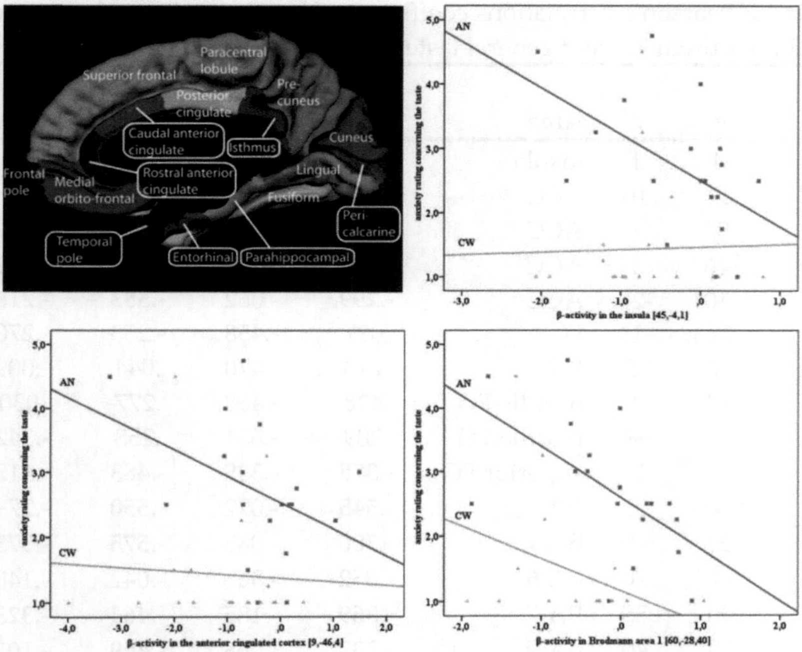

Figure 9: Association between brain activity due to the administration of a taste in general and anxiety ratings of the administration of a taste in the ROIs insula [45,-4,1], anterior cingulate cortex [9,-46,4] and Brodmann area 1 [60,-28,40] for AN patients and CW

Beta-activity in the brain and its associations with anxiety ratings (using a scale from 1 (not at all) to 7 (very much)) in AN patients vs. CW; AN: women with anorexia nervosa, restrictive type (N=20), CW: healthy control women (N=20).

Table 19: Pearson's correlations coefficient of brain activity with stress of fMRI-measurement, and general taste anxiety in AN and CW in the ROIs

				stress		taste - anxiety	
x	**y**	**z**	**Area**	**r (AN)**	**r (CW)**	**r (AN)**	**r (CW)**
45	-4	1	insula	**-.558**	.119	**-.522**	.054
-12	29	16	ACC	**-.348**	-.248	**-.333**	.041
3	-52	7	ACC	**-.623**	-.074	**-.522**	-.109
9	-46	4	ACC	**-.605**	-.049	**-.413**	-.052
0	-10	-20	ACC	**-.299**	-.062	**-.583**	-.216
-24	32	-11	FC	.027	**-.458**	-.253	-.270
-12	47	-8	FG	.213	**-.470**	.044	-.009
-21	47	-8	middle FG	**.428**	**-.468**	.277	.030
27	47	-8	middle FG	.209	**-.631**	.258	**-.342**
18	50	19	superior FG	**-.378**	**-.335**	**-.483**	-.112
42	-34	25	IPL	**-.545**	-.072	**-.550**	-.279
60	-28	40	BA 1	**-.500**	-.088	**-.575**	**-.373**
-21	-1	70	BA 6	-.052	**-.366**	-.042	-.140
3	-61	55	BA 7	**-.369**	-.185	**-.464**	**-.323**
-27	50	40	BA 9	-.233	-.068	**-.449**	-.102
-6	-61	67	precuneus	-.067	-.045	-.256	-.104
3	-43	67	PCG	.070	.056	-.067	.033
3	-1	4	LV	**-.473**	-.200	**-.470**	-.173
9	-1	-14	cerebrum	-.134	-.188	**-.617**	**-.536**
-12	32	7	CC	.060	**-.353**	.002	.026
-3	26	4	CC	.044	**-.407**	-.104	-.082
-6	-37	4	CC	**-.492**	-.061	**-.442**	.088
21	-13	-14	PHG	**-.575**	.133	**-.374**	-.016
24	5	-14	BA 34	**-.463**	.022	**-.368**	-.070

AN: women with anorexia nervosa (N=20), CW: healthy control women (N=20), r=Pearson's correlation coefficient (0.3<r≤0.5: mid-level associations; r>0.5: high-level associations). ACC: anterior cingulate cortex, FC: frontal cortex, FG: frontal gyrus, IPL: inferior parietal lobule, BA: Brodmann area, PCG; postcentral gyrus, LV: lateral ventricle, CC: Corpus callosum, PHG: parahippocampal gyrus.

3.3.6.3 Association between the brain activity due to a taste in general and pleasantness ratings concerning the taste

Our analyses revealed an association between the brain activity due to a taste and the pleasantness rating of the administration of a taste in AN patients in only four ROIs including Brodmann area 7 and 34, the right cerebrum, and the parahippocampal gyrus (table 20). A higher pleasantness rating in AN patients was associated with a higher brain activity in the ROIs. In the CW, an association between these two variables was only found in Brodmann area 1 (table 20). A higher pleasantness rating was again related to a higher brain activity.

3.3.6.4 Association between brain activity due to a taste in general and depression

Our analyses revealed an association between the brain activity due to a taste and the depression score for AN patients in eleven ROIs including the anterior cingulated cortex, the inferior parietal lobe, Brodmann areas 6 and 7, the corpus callosum, and the parahippocampal gyrus (table 20). A higher depression score among AN patients was associated with a lower brain activity in the ROIs. In contrast to this, an association between these two variables was found in six different regions in the CW including the frontal cortex, the right cerebrum, and Brodmann area 34 (table 20). A higher depression score was again related to a lower brain activity. Figure 10 exemplarily shows the association between brain activity due to the administration of a taste in general and the stress rating for AN patients and CW for the ROIs anterior cingulate cortex [9, -46,4], frontal cortex [-24,32,-11], and parahippocampal gyrus [21,-13,-14] .

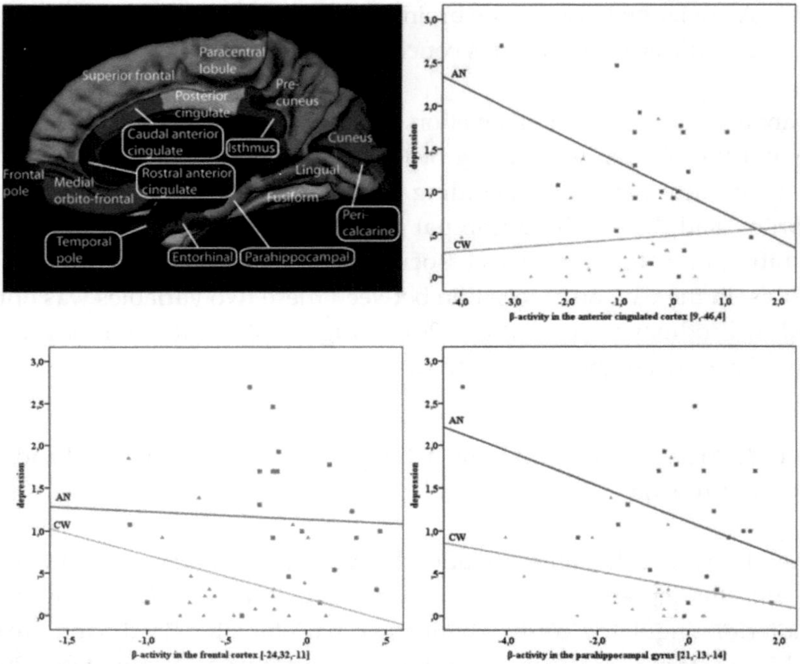

Figure 10: Association between brain activity due to the administration of a taste in general and depression in the ROIs anterior cingulate cortex [9, -46,4] frontal cortex [-24,32,-11], and parahippocampal gyrus [21,-13,-14] for AN patients and CW

Beta-activity in the brain and its associations with depression in AN patients vs. CW; AN: women with anorexia nervosa, restrictive type (N=20), CW: healthy control women (N=20).

Table 20: Pearson's correlations coefficient of brain activity with taste pleasantness ratings and depression in AN patients and CW in the ROIs

				pleasantness		depression	
x	**y**	**z**	**Area**	**r (AN)**	**r (CW)**	**r (AN)**	**r (CW)**
45	-4	1	insula	.272	.189	-.240	-.055
-12	29	16	ACC	.199	.221	**-.518**	-.251
3	-52	7	ACC	.239	-.122	**-.506**	-.098
9	-46	4	ACC	.144	-.244	**-.407**	.091
0	-10	-20	ACC	.135	.111	-.184	**-.417**
-24	32	-11	FC	.083	.216	-.047	**-.336**
-12	47	-8	FG	.034	-.025	-.133	-.272
-21	47	-8	middle FG	-.178	-.067	.072	-.013
27	47	-8	middle FG	-.264	.297	-.144	**-.485**
18	50	19	superior FG	.490	.199	-.195	**-.393**
42	-34	25	IPL	.273	.184	**-.414**	-.195
60	-28	40	BA 1	.269	**.331**	-.192	-.174
-21	-1	70	BA 6	.132	-.192	**.429**	-.128
3	-61	55	BA 7	**.330**	.154	**-.334**	-.281
-27	50	40	BA 9	.715	-.061	.150	-.256
-6	-61	67	precuneus	.228	.130	-.085	-.209
3	-43	67	PCG	.250	.023	.164	-.128
3	-1	4	LV	.248	.233	**-.379**	-.110
9	-1	-14	cerebrum	**.379**	.292	**-.397**	**-.313**
-12	32	7	CC	-.036	.110	**-.304**	-.214
-3	26	4	CC	.048	-.024	-.047	-.267
-6	-37	4	CC	.288	-.238	**-.469**	.153
21	-13	-14	PHG	**.302**	-.036	**-.421**	-.235
24	5	-14	BA 34	**.344**	.273	-.092	**-.323**

AN: women with anorexia nervosa (N=20), CW: healthy control women (N=20), r=Pearson's correlation coefficient ($0.3<r\leq0.5$: mid-level associations; $r>0.5$: high-level associations). ACC: anterior cingulate cortex, FC: frontal cortex, FG: frontal gyrus, IPL: inferior parietal lobule, BA: Brodmann area, PCG; postcentral gyrus, LV: lateral ventricle, CC: Corpus callosum, PHG: parahippocampal gyrus.

3.3.6.5 Association between the brain activity due to a taste in general and anxiety

Our analyses revealed an association between the brain activity due to a taste and the anxiety score in the questionnaire "symptom checklist" (SCL) in AN patients in ten ROIs including the anterior cingulated cortex, the inferior parietal lobe, Brodmann areas 6 and 9, the corpus callosum, and the parahippocampal gyrus (table 21). A higher anxiety score in AN patients was associated with a lower brain activity in the ROIs. In contrast to this, an association between these two variables was found in 13 different regions in the CW including the frontal cortex, the inferior parietal lobe, Brodmann areas 7 and 34, the right cerebrum, and the corpus callosum (table 21). A higher anxiety score was again related to a lower brain activity. Figure 11 exemplarily shows the association between brain activity due to the administration of a taste in general and anxiety for AN patients and CW for the following ROIs: the anterior cingulate cortex [9, -46,4], the frontal cortex [-24,32,-11], and the inferior parietal lobe [42, -34,25] .

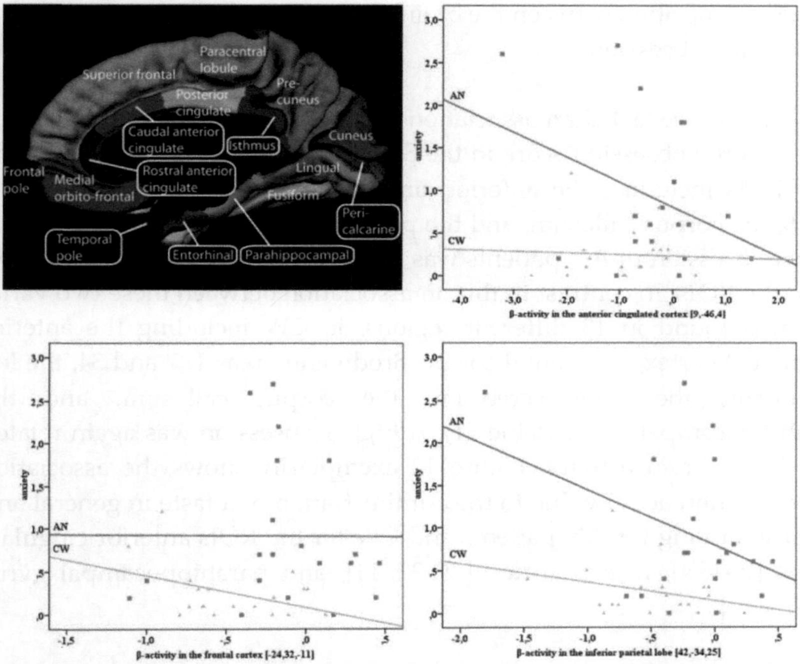

Figure 11: Association between brain activity due to the administration of a taste in general and the anxiety rating in the SCL questionnaire for the ROIs anterior cingulate cortex [9,-46,4] frontal cortex [-24,32,-11], and inferior parietal lobe [42,-34,25] for AN patients and CW

Beta-activity in the brain and its associations with anxiety in AN patients vs. CW; AN: women with anorexia nervosa, restrictive type (N=20), CW: healthy control women (N=20).

3.3.6.6 Association between the brain activity due to a taste in general and obsession

Our analyses revealed an association between the brain activity due to a taste and the obsession score in the SCL questionnaire in AN patients in nine ROIs including the anterior cingulated cortex, Brodmann areas 6 and 9, the corpus callosum, and the parahippocampal gyrus (table 21). A higher obsession in AN patients was associated with a lower brain activity in the ROIs. In contrast to this an association between these two variables was found in 13 different regions, in CW including the anterior cingulated cortex, the frontal cortex, Brodmann areas 1, 7 and 34, the left precuneus, the right cerebrum, the corpus callosum, and the parahippocampal gyrus (table 21). A higher obsession was again related to a lower brain activity. Figure 12 exemplarily shows the association between brain activity due to the administration of a taste in general and the stress rating for AN patients and CW for the ROIs anterior cingulate cortex [9,-46,4], frontal cortex [-24,32,-11], and parahippocampal gyrus [21,-13,-14] .

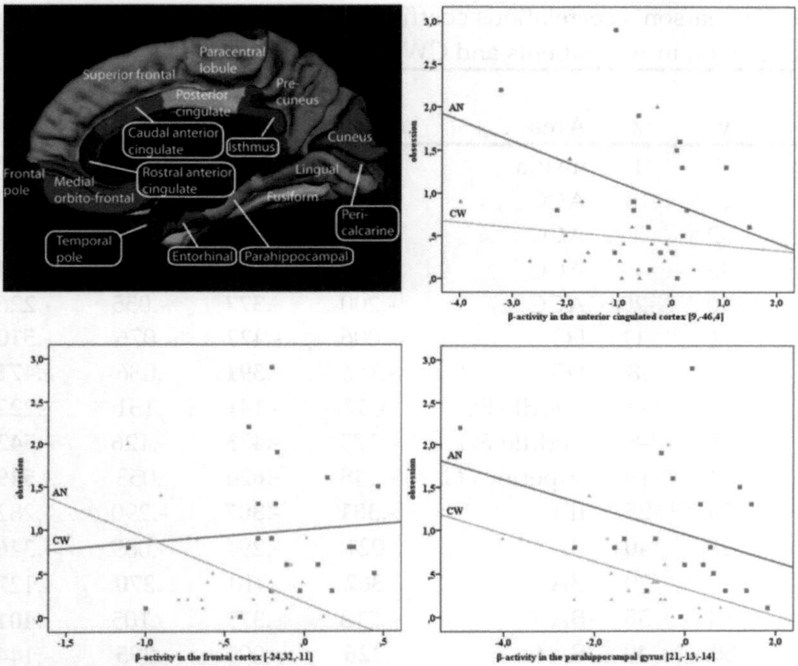

Figure 12: Association between brain activity due to the administration of a taste in general and the obsession rating in the SCL questionnaire for AN patients and CW in the ROIs anterior cingulate cortex [9,-46,4] frontal cortex [-24,32,-11], and parahippocampal gyrus [21,-13,-14]

Beta-activity in the brain and its associations with obsession in AN patients vs. CW; AN: women with anorexia nervosa, restrictive type (N=20), CW: healthy control women (N=20).

Table 21: Pearson's correlations coefficient of brain activity with anxiety, and obsession in AN patients and CW in the ROIs

				anxiety		obsession	
x	y	z	Area	r (AN)	r (CW)	r (AN)	r (CW)
45	-4	1	insula	-.115	-.189	-.109	-.227
-12	29	16	ACC	**-.495**	**-.566**	**-.448**	**-.449**
3	-52	7	ACC	**-.403**	-.264	**-.387**	-.246
9	-46	4	ACC	**-.357**	-.066	**-.328**	-.114
0	-10	-20	ACC	-.200	**-.377**	-.055	-.226
-24	32	-11	FC	-.006	**-.422**	.076	**-.510**
-12	47	-8	FG	-.166	**-.391**	.086	**-.471**
-21	47	-8	middle FG	.057	-.141	.131	-.222
27	47	-8	middle FG	-.173	**-.455**	-.126	**-.542**
18	50	19	superior FG	-.148	**-.624**	.053	**-.549**
42	-34	25	IPL	**-.381**	**-.307**	-.290	-.282
60	-28	40	BA 1	.021	-.294	-.089	**-.346**
-21	-1	70	BA 6	**.362**	.010	**.370**	-.127
3	-61	55	BA 7	-.213	**-.373**	-.105	**-.401**
-27	50	40	BA 9	**.326**	-.099	**.395**	-.144
-6	-61	67	precuneus	-.035	**-.321**	.059	**-.325**
3	-43	67	PCG	.178	-.230	**.331**	-.163
3	-1	4	LV	**-.349**	**-.411**	**-.338**	**-.379**
9	-1	-14	cerebrum	**-.354**	**-.520**	-.188	**-.456**
-12	32	7	CC	-.281	**-.395**	-.242	**-.339**
-3	26	4	CC	-.107	-.208	-.011	-.296
-6	-37	4	CC	**-.490**	-.090	**-.432**	-.019
21	-13	-14	PHG	**-.318**	-.252	**-.332**	**-.391**
24	5	-14	BA 34	-.058	**-.620**	.021	**-.539**

AN: women with anorexia nervosa (N=20), CW: healthy control women (N=20), r=Pearson's correlation coefficient ($0.3<r\leq0.5$: mid-level associations; $r>0.5$: high-level associations). ACC: anterior cingulate cortex, FC: frontal cortex, FG: frontal gyrus, IPL: inferior parietal lobule, BA: Brodmann area, PCG; postcentral gyrus, LV: lateral ventricle, CC: Corpus callosum, PHG: parahippocampal gyrus.

3.3.6.7 Association between the brain activity due to a taste in general and obsessional control

Our analyses revealed an association between the brain activity due to a taste and obsessional control in the questionnaire "Hamburger Obsessive-Compulsive Behavior Inventory" (HZI) in AN patients in eleven ROIs including the insula, the anterior cingulated cortex, the frontal cortex, Brodmann area 7, and the corpus callosum (table 22). A higher obsessional control in AN patients was associated with a lower brain activity in the ROIs. In contrast to this, an association between these two variables was found in eight different regions in CW including the insula, the anterior cingulated cortex, the frontal cortex, Brodmann areas 1 and 34, the corpus callosum, and the parahippocampal gyrus (table 22). A higher obsessional control was again related to a lower brain activity. Figure 13 exemplarily shows the association between brain activity due to the administration of a taste in general and obsessional control for AN patients and CW for the ROIs insula [45,-4,1], anterior cingulate cortex [9,-46,4], and frontal cortex [-24,32,-11].

3.3.6.8 Association between the brain activity due to a taste in general and the duration of the illness

Our analyses revealed an association between the brain activity due to a taste and the duration of suffering from AN in the patients with anorexia in twelve ROIs including the anterior cingulated cortex, the frontal gyrus, the inferior parietal lobe, Brodmann area 7, the right cerebrum, the corpus callosum, and the parahippocampal gyrus (table 22). A longer duration of the illness in AN patients was associated with a lower brain activity in the ROIs. Figure 14 exemplarily shows the association between brain activity due to the administration of a taste in general and the duration of the illness for AN patients and CW for the ROIs anterior cingulate cortex [9,-46,4], frontal cortex [-24,32,-11], and parahippocampal gyrus [21,-13,-14].

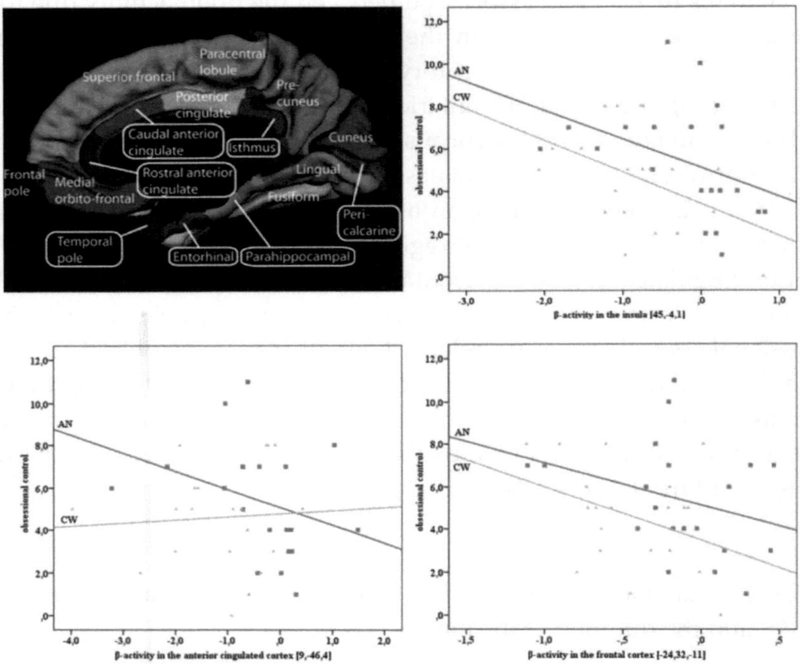

Figure 13: Association between brain activity due to the administration of a taste in general with obsessional control in the HZI questionnaire for AN patients and CW in the ROIs insula [45,-4,1], anterior cingulate cortex [9,-46,4], and frontal cortex [-24,32,-11]

Beta-activity in the brain and its associations with obsessional control in AN patients vs. CW; AN: women with anorexia nervosa, restrictive type (N=20), CW: healthy control women (N=20).

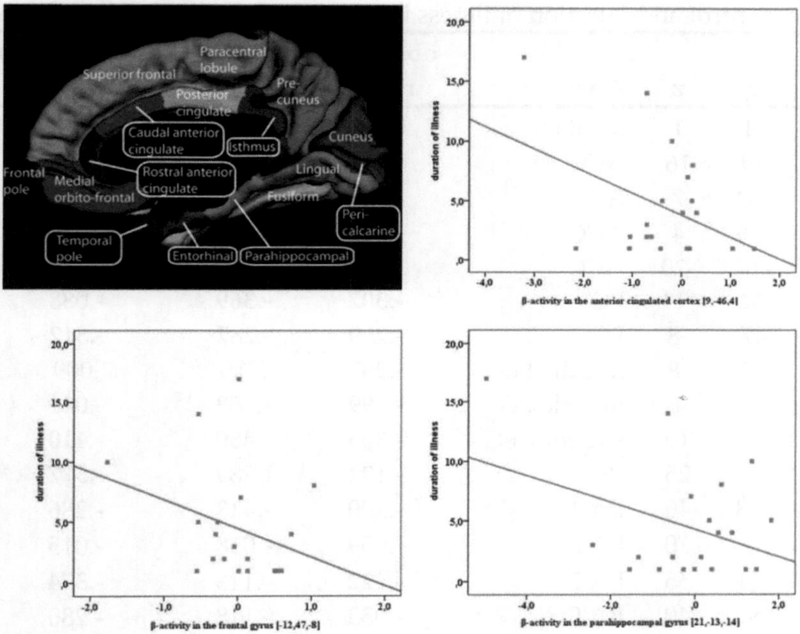

Figure 14: Association between brain activity due to the administration of a taste in general with the duration of the illness for AN patient in the ROIs anterior cingulate cortex [9,-46,4] frontal gyrus [-12,47,-8], and parahippocampal gyrus [21,-13,-14]

Beta-activity in the brain and its associations with the duration of the illness in years in AN patients; AN: women with anorexia nervosa, restrictive type (N=20).

Table 22: Pearson's correlations coefficient of brain activity with obsessional control and duration of illness in AN patients and CW in the ROIs

				obsessional control		illness duration
x	y	z	Area	r (AN)	r (CW)	r (AN)
45	-4	1	insula	**-.394**	**-.434**	-.213
-12	29	16	ACC	**-.538**	**-.434**	**-.348**
3	-52	7	ACC	**-.358**	-.061	**-.538**
9	-46	4	ACC	**-.329**	.065	**-.414**
0	-10	-20	ACC	-.162	.080	**-.633**
-24	32	-11	FC	**-.307**	**-.369**	-.088
-12	47	-8	FG	-.259	-.287	**-.312**
-21	47	-8	middle FG	-.297	.010	.009
27	47	-8	middle FG	**-.299**	-.258	-.041
18	50	19	superior FG	**-.305**	**-.480**	**-.410**
42	-34	25	IPL	-.124	-.289	**-.577**
60	-28	40	BA 1	-.200	**-.413**	-.286
-21	-1	70	BA 6	-.059	-.058	-.015
3	-61	55	BA 7	**-.422**	-.113	**-.374**
-27	50	40	BA 9	-.033	-.118	-.280
-6	-61	67	precuneus	-.215	-.035	-.189
3	-43	67	PCG	-.080	-.160	-.082
3	-1	4	LV	**-.371**	-.252	**-.343**
9	-1	-14	cerebrum	-.032	-.179	**-.474**
-12	32	7	CC	**-.561**	**-.318**	-.054
-3	26	4	CC	-.283	-.149	-.172
-6	-37	4	CC	**-.478**	.165	**-.435**
21	-13	-14	PHG	-.166	**-.389**	**-.381**
24	5	-14	BA 34	-.279	**-.436**	-.196

AN: women with anorexia nervosa (N=20), CW: healthy control women (N=20), r=Pearson's correlation coefficient ($0.3<r\leq0.5$: mid-level associations; $r>0.5$: high-level associations). ACC: anterior cingulate cortex, FC: frontal cortex, FG: frontal gyrus, IPL: inferior parietal lobule, BA: Brodmann area, PCG; postcentral gyrus, LV: lateral ventricle, CC: Corpus callosum, PHG: parahippocampal gyrus.

4 Discussion

4.1 Psychological aspects of anorexia nervosa

The aim of this study was the analysis of psychological and neurobiological aspects (structural and functional alterations concerning the processing of taste stimuli) of AN. All AN patients in this study suffered from *medical complications* such as low blood pressure (hypotension), amenorrhoea, a low heart beat (bradycardia), electrolyte disturbances, and diarrhea. This indicates a high health risk of the patients with AN as the medical complications which are related to the disease can lead to mortality (Lemberg, 1999).

Some of the patients in this study declared in their anamnesis that they suffered from the common *co-morbid disorders*: anxiety disorder, obsessive-compulsive disorder and depression. In addition to this, the results of the questionnaire asking for medical complaints in this study showed that patients with AN somatize more frequently than CW, are more uncertain in social situations and suffer more often from mood disorder, phobia, paranoid thinking, and psychoticism. One-third of the patients also took medication (an SSRI) for their depressive symptoms, although the results of studies analyzing the efficiency of psychopharmacological treatment in AN patients are sparsely encouraging (Lahousen et al., 2003). Additionally, the results of the questionnaire asking for co-morbid symptoms confirm that the patients with AN in this study were more often depressive, had higher scores concerning their state and trait anxiety and had a higher overall obsessive-compulsive score than the CW – they especially differed from the CW concerning organizing in an obsessive way. This is in line with previous studies which showed that individuals with AN have higher scores in questionnaires asking for symptoms of obsession (Kaye et al., 2005, 2009; Lemberg, 1999; Uher et al., 2003), rigidity, anxiety and depression (Barbarich et al., 2003; Kaye et al., 2005, 2009; Lemberg, 1999; Tejado et al., 2010; Watkins, 2011). It still remains unclear whether subjects suffering from an ED have a primary disturbance that modulates feeding or whether a disturbed appetite is secondary to the co-morbid symptoms such as anxiety and obsession (Wagner et al., 2008).

Overall, the number of patients who rated their *subjective health* as not good was rather high. One reason for this could be that all of the patients were currently in treatment, probably due to the degree of suffering from AN. Additionally, the results of this study evidenced that the *quality of life* is impaired in patients with AN. The patients in this study reported to have a worse quality of life compared to the CW regarding physical and psychological health and concerning environmental factors. Moreover, nearly one third of the patients in this study have already *self-harmed,* which corresponds to the results of previous studies reporting a prevalence of self-injurious behavior between 25 % and 45 % of patients who suffer from an ED (Eberly, 2005).

Nearly one third of the patients with AN in this study had already thought about committing *suicide* and nearly one fifth had already made a suicide attempt. Compared to the results of previous studies which reported a prevalence of suicide attempts between 3 % and 20 % in patients with AN (Franko & Keel, 2006; Milos et al., 2004), the number of individuals with AN who had already attempted suicide in this study is rather high. This high prevalence of suicide attempts in patients with AN in this study is quiet scary considering the fact that suicide attempts are a risk factor for later completed suicide (Bulik et al., 1999; Foulon et al., 2007; Herzog et al., 2000; Keel et al., 2003; Milos et al., 2004; Pryor et al., 1996; Youssef et al., 2004) and that suicide is one of the major causes of death leading to the high mortality rate of patients with AN (Bulik et al., 1999; Lemberg, 1999). This disease has the highest mortality rate of all mental disorders (Bulik et al., 1999; Frank et al., 2008) – about 50 % of individuals with AN develop a chronic illness or die (Kaye et al., 2009) – which demonstrates the need for a strong public health program to prevent the disease and associated co-morbid symptoms as well as long-term effects.

4.1.1 *Personality*

Specific personality traits which precede the onset of an ED and persist after recovery can lead to a predisposition to develop an ED (Kaye et al., 1998, 2005, 2009, 2014; Lemberg, 1999). The results of this study show that patients with AN have a lower life satisfaction, are more inhibited and emotional, suffer from more strain and a higher number of physical complaints and additionally, they rate themselves as less extroverted. The patients with AN in this study differed from the CW in the subscales of the EDI. They had a higher score concerning drive for thinness, ineffectivity, perfectionism, interpersonal distrust, interceptive awareness, maturity features, asceticism, and social insecurity. This corresponds to the results of other studies which showed that patients suffering from anorexia show specific personality traits including perfectionism, conformity, constriction of affect and emotional expressiveness, feelings of ineffectiveness, high self-control, social introversion, overly compliant behavior, and reduced social spontaneity (Barbarich et al., 2003; Kaye et al., 2013; Klump et al., 2000; Lemberg, 1999). Kaye et al. (2013) suggest that the specific personality characteristics might be related to the exaggerated serotonin activity which diminishes effects of the dopamine reward systems. In addition to this, it is possible that an increased activity in certain brain regions (such as the frontal cortex or the dorsal striatum) might contribute to the characteristics which are associated with the disease such as perfectionism or strategizing (Kaye et al., 2013). The results of this study evidenced that patients with AN have specific personality characteristics which should be targeted in therapy.

4.1.2 *Stress and coping*

The following discussion regarding differences in perceived stress and the use of coping strategies are withdrawn from a research paper by Burkert et al. (2015) and were adapted for this dissertation.

The results of this study evidenced deficiencies among AN patients as to the use of positive coping strategies. Compared to their controls, AN patients made less frequent use of positive coping strategies such as deemphasizing, situation control, positive self-instructions, and call for support. This is in accordance with the results of other studies showing that anorexia is associated with dysfunctional coping strategies (Bloks et al., 2004; Brytek, 2006; Nakahara et al., 2000; Taylor & Stanton, 2007; Troop et al., 1998). Prior studies showed AN patients failing to cope positively with stress (Brytek, 2006) and less often using coping styles such as emotional support, positive self-instructions, and positive reframing (Brytek-Matera and Schlitz, 2013). Moreover, studies were able to show that patients with AN cope with feelings through their eating behavior: sadness and fear, for example, are managed through restrictive eating and purging, anger is managed through self-control, self-harm and exercising (Espeset et al., 2012). Additionally, determined approaches to solve problems, seeking social support (Villa et al., 2009), cognitive avoidance, cognitive rumination or deemphasizing (Troop et al., 1998) seem to be impaired in patients with AN. Swanson et al. (2010) report that patients with AN more often use negative coping styles such as avoidance as a mechanism to manage stress. Other studies report that in AN patients, coping styles such as positive attitude, planning and social support seem to be impaired (Villa et al., 2009).

Moreover, the results of this study show that AN patients perceived more stress during the MRI measurement. This might be related to the smaller observed volumes of hippocampal sub-structures, as AN is associated with a higher cortisol secretion (Seed et al., 2000). As a consequence, cortisol effects might induce a higher stress level as the hippocampus has a high density of glucocorticoid receptors and the maintenance of AN is exacerbated by stress (Jappe et al., 2014).

4.1.3 *Body image*

The results of this study show that body perception is disturbed in patients with AN. The patients rated themselves as less attractive than the CW, were more uncertain and concerned about their body and felt more physical-sexual discomfort. This corresponds with the results of previous studies showing that the disease is associated with body image distortion (Barbarich et al., 2003; Ehrig, 2003; Espinoza et al., 2009; Kaye et al., 2005; Lemberg, 1999; Robertson, 2009; Tejado et al., 2010; Watkins, 2011). Moreover, there are studies which found that deficiencies in body perception seem to persist after recovery (Golden & Meyer, 2004). Neurobiological studies suggest that in patients with AN, deficient brain connectivity might contribute to visuspatial difficulties. As a consequence, the integration of visual and somatosensory information might be impaired and in turn contributes to the disturbed body image which is present in individuals with AN (Favaro et al., 2012). George et al. (2011) found that patients with AN overestimate body sizes and show a stronger eye fixation pattern on body parts such as hip and collar bones. This might alter their judgement of an ideal body size. In addition to this, alterations in brain activity, e.g. in the parietal cortex, might have an influence on spatial orientation disturbances and subsequently on body image distortions, which are present in individuals suffering from anorexia (Guardia et al., 2012).

4.2 Neurobiological aspects of anorexia nervosa

4.2.1 Structural changes

The results of this study show that patients with AN do not differ from CW in their total intracranial volume or cortical white matter volume. In contrast to this, Frank et al. (2013) found that patients with AN have a reduced white matter volume compared to controls. The results of this study showed a significantly reduced gray matter volume and less overall cortex volume in patients with AN compared to healthy controls. This result is in line with other studies which evidenced a decreased overall brain mass in patients suffering from AN (Katzman et al., 1996; Ellison et al., 1998; Frank et al., 2004; Tejado et al., 2010). Unfortunately, as patients who were currently affected with the disease were analyzed in this study, it was not possible to differentiate between the primary disorder and the secondary phenomena accompanying the disease.

Prior MRI studies found a localized reduction in gray matter volume in AN patients in several brain regions including parietal and temporal areas (Castro-Fornieles et al., 2007), occipital and medial frontal brain areas as well as structures such as the anterior cingulated cortex (Friedrich et al., 2012; Mainz et al., 2012; Muehlau et al., 2007), the amygdala, the putamen (Friedrich et al., 2012; Mainz et al., 2012), the cerebellum (Miwa et al., 2004), the middle cingulated cortex, the precuneus (Gaudio et al., 2011), the parahippocampal gyrus, the left cerebellum, the left fusiform gyrus, and the right posterior cingulated cortex (Brooks et al., 2011). Brain regions which were especially affected in patients in this study were the corpus callosum, the thalamus, the choroid plexus, and the amygdala. All these regions were significantly reduced in patients with AN compared to CW, except for the left choroid plexus which was enlarged in patients with AN. The reduced volume of the thalamus being responsible for affect regulation (Mertens et al, 2014) might be associated with emotional deficiencies. As a ventral neurocircuit, which includes the amygdala, is responsible for identifying the emotional aspects of stimuli and generating an affective response (Phillips et al., 2003), it is possible that deficiencies in emotional regulation in patients with AN are related to changes in neurocircuits. In addition to this, the thalamus, together with the insula, is responsible for integrating signals related to hunger and satiety and also caloric content (Van Rijn et al., 2014). The results of this study show a reduced volume of the thalamus in AN patients, which might have an influence on the patients' food intake.

Prior studies have shown that in patients suffering from AN, deficits in the integrity of limbic and associational regions exist (Kazlouski et al., 2011). Studies also showed a reduced cerebral blood flow and global hypometabolism in AN patients (Delvenne et al., 1995, 1996, 1997; Kerem & Katzman, 2003; Tejado et al., 2010), as well as alterations concerning neurotransmitter levels (Bailer et al., 2006; Barbarich et al., 2003; Kaye et al., 2005; Tejado et al., 2010). As these functions were not analyzed in this study, it was not possible to either verify or falsify any of this evidence.

Whether the structural and functional abnormalities are fully reversible is still not evident. The results of prior studies had controversial results (Fuglset & Frampton, 2011; Frampton et al., 2011; Giordano et al., 2001; Golden & Meyer, 2004; Kaye et al., 2011; Mainz et al., 2012; Wagner et al., 2006a). In this study, structural changes of the brain were analyzed in patients currently affected by the disease. Considering the fact that malnutrition causes changes in neuroendocrine function, further longitudinal studies in recovered AN patients are required.

4.2.2 *Hippocampal changes and associations with psychological symptoms in AN patients*

The following discussion regarding changes concerning the volume of the hippocampus and its associations with other brain areas, stress and coping are withdrawn from a research paper by Burkert et al. (2015) and was adapted for this dissertation.

The results of this study evidence smaller volumes for certain hippocampal sub-structures while the hippocampal fissure was enlarged; the results also show an association between the volume of the hippocampus and the use of positive coping strategies as well as stress levels.

Prior studies have analyzed the hippocampal volume in patients suffering from AN. Those studies showed that AN patients had a smaller overall hippocampal volume (Connan et al., 2006) and a smaller right and left hippocampus-amygdala formation (Giordano et al., 2001), even after recovery. No difference between the AN and CW groups was found in the overall volume of the hippocampus in that study. However, in this study such a difference was found in certain hippocampal sub-structures. The AN patients had a smaller presubiculum, a tinier fimbria, and larger fissure compared to the CW. As restrictions concerning the integrity of limbic and associational regions were observed in AN patients (Kazlouski et al., 2011) and the hippocampus is part of the limbic system, which is responsible for emotion regulation (Andersen et al., 2006), deficits in coping and emotional distress among AN patients might at least partly be explained by alterations in the hippocampal areas.

The results of this study showed a reduction in the hippocampal volume to be related to higher stress levels and a higher use of positive coping in AN patients. This higher use of positive coping might at least in part be explained by the fact that AN patients who experienced more stress showed a higher use of positive coping. However, as this association contradicts the finding that AN is associated with lesser use of positive coping strategies, it should be analyzed in detail in further studies.

Associations between the hippocampal volume and stress or coping were also found in other mental disorders. Impairments in the use of positive coping strategies were, for example, found in policemen after traumatic events (Henning-Fast et al., 2009). A reduction of hippocampal volume was also observed in patients with borderline personality disorder (Rodrigues et al., 2011), dissociative identity disorder (Vermetten et al., 2006), schizophrenia, and depression (Campbell & Macquenn, 2004; Rodrigues et al., 2011). In addition to this, animal models showed chronic unpredictable stress to induce depressive-like behaviors, passive behavioral coping or anhedonia and to have an effect on hippocampal neurogenesis (Mahar et al., 2014). According to Barr and Foster (2011), the reduced serotonergic activity in the hippocampus might lead to altered stress responses and associated coping processes in rats. Other studies showed that anti-depressant medication can increase the hippocampal volume (Heechul et al., 2012). As evidence exists that in females with AN disturbances of serotonin pathways play a role in the pathogenesis and pathophysiology of the disease (Bailer & Kaye, 2010), these disturbances might have an influence on coping processes.

Stress seems to induce an atrophy of dendrites in pyramidal cells of the CA3 region in the hippocampus; additionally, stress inhibits the genesis of new neurons in the dentate gyrus (Andersen et al., 2006). Especially in patients suffering from posttraumatic stress disorder, a selective volume loss was observed in the hippocampal subfields of the CA3 region (Wang et al., 2010). The results of this study did not show a reduced volume in these regions in patients with AN, but evidenced a smaller presubiculum and fimbria as well as a larger fissure. The presubiculum is located at the boundary between the hippocampus and the cortex (Simonnet et al., 2013), is highly interconnected with subcortical structures (Lavenex and Lavenex, 2013) and is involved in the processing of spatial information (Simonnet et al., 2013) and orientation (Andersen et al., 2006). The fimbria-fornix region of the brain is also highly interconnected with other regions. Animal studies showed that lesions in this brain area are associated with impairments of spatial learning and memory capabilities and lead to cognitive impairment (Feldman et al., 2010). Several cognitive impairments concerning cognitive functions were found in patients suffering from AN in prior studies, especially regarding learning, memory, executive functioning, and visuospatial abilities (Fowler et al., 2006; Weider et al., 2014). This study showed that the presubiculum and the fimbria, which are regions that are key to spatial information processing, were reduced in females suffering from AN. As body distortion is a core phenomenon of the disease and has also been related to spatial orientation disturbances (Guardia et al.,

2012), the disturbed body image in AN patients might also, at least in part, be associated with the hippocampal presubiculum and the fimbria and their interconnection with other brain areas which are involved in spatial processing.

4.2.3 *Brain activity as a response to a taste in general*

The results of the whole brain analyses of this study show that disturbances concerning the processing of taste stimuli exist in the relevant brain regions which are responsible for reward, decision-making, emotional responses, and control. Especially in the insula, the anterior cingulated cortex, the frontal cortex, the inferior parietal lobe, and the parahippocampal gyrus alterations in brain activity as a response to the administration of a taste were found in patients suffering from AN compared to healthy controls. This is in line with prior fMRI studies suggesting that brain activity in systems which are responsible for interoception, reward, emotionality, and inhibition is disturbed in patients with AN and might therefore contribute to food avoidance (Berthoud, 2011; Kaye et al., 2011, 2013). As the results of this study showed an altered insula response due to the administration of a taste, and as the insula is responsible for integrating gustatory stimuli with interoceptive signals (Craig, 2009; Kaye et al., 2009; Nagai et al., 2007; Oberndorfer et al., 2013), this might lead to the inability to react positively with reward to taste stimuli. As a consequence, this might at least in part contribute to the restrictive eating behavior.

The amygdala, the ventral anterior cingulated cortex (ACC) and the orbifrontal cortex (OFC) are interconnected with the insula. Insula inputs to the striatum are suggested to mediate behavior including eating, and are responsible for information processing concerning the physiological condition of the body, the incentive value of food and, in turn, the integration of interoceptive stimuli (Wagner et al., 2008). This study did not only find an alteration of taste processing in the insula, but also in the ACC and the frontal cortex. As pleasure, motivation, and cognitive aspects of tastes are modulated in these areas, the altered response might lead to the decision of not to eat in patients with AN. Additionally, as the frontal areas in the brain are involved in inhibition, decision-making, execution of behavior, and reward-seeking (Geliebter et al., 2006), the altered brain activity in AN patients might influence the decision not to eat. Kaye et al. (2009) hypothesized that in AN patients, the reward response might be mediated by alterations concerning information processing towards a strategic approach, thus leading to the satisfaction of long-term goals (lose weight and avoid food) rather than short-term goals (allay hunger). Furthermore, as a higher anterior cingulate cortex and medial prefrontal cortex activity was observed in both ill and recovered individuals with AN when they were confronted with food, it might be a trait marker for AN (Frank et al., 2004; Uher et al., 2003, 2004). Additionally, also structural changes of the brain in patients suffering from AN seem to be associated with taste pleasantness and reward of taste stimuli (Frank et al., 2013). In this study, patients suffering from AN showed a reduced volume of the thalamus, which might also have an influence on caloric intake in the patients (Van Rijn et al., 2014).

Evidence exists concerning the parahippocampal gyrus being implicated in the processing of emotions, especially when AN patients are confronted with their own body. Alterations in brain activity in the insula, the dorso-lateral prefrontal cortex, the cerebellum, and the amygdala have also been related to alterations concerning body image perceptions in AN (Tejado et al., 2010). The results of this study showed a higher brain activity in some of these brain regions (the insula, the frontal cortex, and the parahippocampal gyrus) as a response to the administration of a taste in general in patients suffering from AN, which might be associated with the symptom of a distorted body image which was also found in the AN patients in this study.

4.2.4 *Brain activity as a response to sucrose, umami, and citric acid in regions of interest*

The primary taste cortex (anterior ventral insula) responds to taste stimuli irrespectively of their modality (Rudenga et al., 2010), and also the OFC is activated by a pleasant as well as an unpleasant taste (O'Doherty et al., 2001). The connectivity, however, between the insula and the associated brain areas (hypothalamus, ventral pallidum, and striatum) seems to depend on the modality of the taste stimuli (Rudenga et al., 2010). Prior studies analyzed the brain activity in a salty or umami taste, but only in healthy individuals (McCabe and Rolls, 2007; Nakamura et al., 2011; Rudenga et al., 2010; Spetter et al., 2010). This is the first study to analyze the brain activity as a response to umami and citric acid in patients suffering from AN.

By analyzing the brain activity as a response to the three different tastes (sucrose, umami, and citric acid), the results of this study showed that the main areas of taste processing (insula, ACC, and frontal cortex) respond to each taste, independently of its flavour. The results of previous fMRI studies which analyzed the brain activity as a response to taste stimuli in patients with AN showed a higher activity in the ventral striatum as a response to a pleasant chocolate taste and in the insula and putamen as a response to an aversive taste (Cowdrey et al., 2011). In addition to this, an altered activity in the insula, the OFC, the ACC and in the mesial temporal and parietal regions was found in AN patients as a response to sucrose (Kaye et al., 2009). It is possible that an activation in the mesial temporal cortex leads to increased anxiety in patients who suffer from AN by an alteration of the amygdala (Frank et al., 2004).

A reduced mean signal change over time was found in the insula and in related brain regions (e.g. the ACC) to the administration of a sucrose taste solution (Oberndorfer et al., 2013; Wagner et al., 2008). In addition to this, studies found a higher beta-activity in the striatum, the insula, the OFC and the amygdala in individuals suffering from AN compared to obese individuals (Frank et al., 2012). Radeloff et al. (2014) reported a reduced response to a high-fat cream and non-caloric viscous stimulus in AN patients compared to individuals suffering from bulimia. Overall, BN seems to be related to potential hypofunctioning of the brain reward system, which may cause these individuals to binge eat in order to compensate for this reward deficit (Bohon & Stice, 2010; Frank et al., 2006; Kaye et al., 2001), while AN is associated with hyperfunctioning, which might cause individuals suffering from the disease to starve themselves.

Overall, the administration of taste stimuli seems to be associated with a higher activity in reward systems which are associated with dopamine release in AN (Frank et al., 2012). Studies evidenced that exaggerated serotonin activity might diminish effects of the dopamine appetite reward systems (Kaye et al., 2013). This indicates that not only disturbances of brain activity in regions which are responsible for food reward seem to have an influence on the disturbed eating behavior in individuals with AN, but that also hormonal changes such as disturbances in the serotonin and dopamine system seem to have an influence on its maintenance (Kaye et al., 2011).

Studies evidenced that after recovery an alteration concerning cognitive taste-processing seems to persist (Castro-Fornieles et al., 2010; Cowdrey et al., 2011; Hay & Sachdev, 2011; Oberndorfer et al., 2013; Wagner et al., 2008). Cognitive-emotive deficits therefore seem to represent a putative endophenotype of AN and suggest a genetic basis for abnormal emotive processing in the disease (Holliday et al., 2005; Roberts et al., 2010; Tencono et al., 2010). Evidence emerging from functional neuroimaging studies raises the possibility of a biological abnormality that may predispose to AN. However, the answer to the question of whether brain activity disturbances are traits of scarves is still unknown. Therefore, longitudinal studies are necessary to analyze whether taste processing is disturbed in the long-run.

4.2.5 *Associations between the brain activity as a response to a taste and psychological factors*

The results of this study evidence that perceived stress seems to be associated with the brain activity in the ROIs due to the administration of a taste in general in AN. In addition to this, the results of this study show that anxiety concerning the administration of a taste influences the brain activity in patients with AN. Brain activity in the frontal cortex has been reported to be responsible for encoding the emotional value of a stimulus (Uher et al., 2003, 2004). Anxiety might therefore alter affective responses in AN patients, leading to increased cognitive rigidity and, subsequently, might have an influence on daily food intake in these patients. Additionally, studies evidenced that patients with AN, and even recovered individuals, have an enhanced ability to pay attention to detail rather than to see the global picture (Kaye et al., 2011; Lopez et al., 2008). This might possibly lead to the fact that anxious patients respond cognitively to stimuli without recognizing the emotional aspects (Kaye et al., 2011). This corresponds with the results postulated in the cognitive emotional theory which says that emotions are caused by cognitive appraisal (Arnold, 1960; Lazarus, 1991). Individuals suffering from AN might therefore be unable to appreciate rewarding effects (Kaye et al., 2011), as the processing of food also includes the internal evaluation of the ingested seen or smelled nutrients (Frank et al., 2013).

Prior studies showed that food ratings have an influence on brain activity. The consumption of palatable foods has been reported to be associated with a greater activation in the OFC and the insula in healthy subjects (Liu et al., 2000; O'Doherty et al., 2001, 2002; Small et al., 2001, 2003, 2008; Volkow et al., 2003; Wagner et al., 2008). The results of our study, however, did not evidence a possible association between the pleasantness ratings of a taste and the brain activity, neither in AN patients nor in CW.

The results of this study show that an association between the comorbid disorders depression, anxiety disorder, obsession and obsessional control with the brain activity due to the administration of a taste in the

ROIs ACC, Brodmann area 6, lateral ventricle, corpus callosum, and parahippocampal gyrus exists in patients with AN. The ACC has been reported to be involved in emotional processing and the integration of interoceptive stimuli (Kaye et al., 2009; Wagner et al., 2008). The altered response in this area in patients with AN might therefore have an influence on their decision not to eat. In addition to this, alterations in the parahippocampal gyrus, as being part of the limbic system, might influence caloric intake in the disease. As alterations of the lateral ventricle were reported to be evident in psychiatric disorders (Van Erp et al., 2014), this might have an influence on the disturbed processing in AN patients.

A potential limitation of this study is the fact that patients with AN currently affected with the disease, who also differed from CW in the co-morbid symptoms were analyzed. Other studies evidenced abnormalities in the neural representation of rewarding stimuli in frontal brain areas in patients suffering from or being at risk for depression (McCabe et al., 2009, 2012; Simmons & Drevets, 2012). Changes in the orbifrontal cortex also seem to exist in patients suffering from obsessive-compulsive disorder (Zald & Kim, 1996). As a consequence, it cannot be stated whether the differences in taste processing are only due to the ED or whether they are (also) caused by other symptoms.

Additionally, the results of this study show that a relation exists between the brain activity in the ROIs due to the administration of a taste and the duration of the illness in patients with AN. Brain activity in regions such as the ACC, the frontal and parahipocampal gyrus decreases with a longer duration of the illness. This indicates that the observed neurobiological changes are associated with long-term effects of the disease. Longitudinal studies are needed, however, to examine which changes precede the onset of the disease and which persist after recovery in order to identify traits which comprise the susceptibility to develop the disease.

4.3 A critical reflection of the current study

According to the biopsychosocial model of anorexia, certain biological (e.g. genetic factors, neurotransmitter levels), psychological (e.g. family factors, personality characteristics), and sociocultural factors (e.g. the importance of weight and shape) have been investigated by researchers and attributed as influencing factors on the development of the disease (Duvvuri & Kaye, 2009; Lemberg, 1999; Lucas, 1981; Polivy & Herman, 2002; Watkins, 2011; Wood, 2011). Overall, weight concerns and the sociocultural importance of body shape are more present in Western countries than in Eastern regions (Katzman & Golden, 2009). It has been discussed that the social pressure to be thin, together with changes during puberty such as gonadal steroids, development, and stress and the presence of certain personality characteristics, feelings of ineffectiveness and loss of control leads to neurobiological changes which enhance dieting, rigidity, obsession, anxiety and depression (Katzman & Golden, 2009; Kaye et al., 2009). In turn, weight loss leads to further dieting and a preoccupation with weight and shape which enhances the desire to lose more weight and starts a vicious cycle (Katzman & Golden, 2009). In fact, while many biological, physiological, and psychological aspects of AN have been attributed to its development (Kaye et al., 1998, 2005, 2009, 2013, 2014; Lemberg, 1999; Nunn et al., 2008), they can also be seen as a consequence of malnutrition and weight loss (Katzman & Golden, 2009).

The results of the present study show that psychological differences in patients with AN exist compared to their normal controls (regarding certain personality characteristics, body image, and coping strategies). Additionally, our study was able to show that also biological changes exist in patients suffering from the disease such as a selective volume loss in special brain areas and an altered taste processing in brain areas which are relevant for reward, and emotional responses. The design of the present study, however, does not allow us to make any assumptions regarding the development of AN in terms of etiologic factors. As we analyzed psychological and neurobiological factors in patients who were currently affected with the disease, we were not able to differentiate between causal and consequential factors. In addition to this, we analyzed both aspects separately. Prior studies evidenced that malnutrition causes changes in neuroendocrine, but also in psychological functioning (Katzman & Golden, 2009); however, these changes have also been attributed to the development of the disease (Kaye et al., 1998, 2005, 2009, 2013, 2014; Lemberg, 1999). One example are the results of the numerous studies which report that certain personality characteristics such as perfectionism, high anxiety, and feeling of ineffectiveness were found to precede the onset of the disease (Katzman & Golden, 2009; Kaye et al., 2009).

Overall, the results of the numerous studies are inconsistent. On the one hand, they show that certain biopsychosocial aspects precede the onset of the disease and certain biopsychosocial aspects go along with the disease. On the other hand, it remains unclear which factors are a cause and which are a consequence. A genetic predisposition to develop an ED, for example, has been discussed based on the elevated risks to develop an ED in family members (Bulik et al., 2006; Strober et al., 2000). But as these families have the same co-socialisation (Kendler & Gardener, 1998; Thornton et al., 2011) and also share common aspects such as a preoccupation with weight and shape and a suppression of emotions (Lemberg, 1999), the development of the disease can also be attributed to familial factors. In addition to this, traumatic influences can lead to both the development of an ED (Johnson et al., 2002) and the inducing of neurobiological changes (Andersen et al., 2006; Apfel et al., 2011; Brown & Morey, 2012; Childress et al., 2013; Dannlowski et al., 2012; Ding et al., 2013; Hall et al., 2012; Lindauer et al., 2005; Papagni et al., 2011; Rocha-Rego et al., 2012; Wang et al., 2010).

To sum up, this study evidences psychological and neurobiological changes in patients with AN, but the present findings do not allow to draw any inferences regarding the role of biological, psychological, and social factors being a cause or a consequence of the disease.

4.4 Strengths and limitations

Among the strengths of this study are the unique database, the matching of females with AN and healthy controls based on their age, and the relatively large sample of individuals who were tested with MRI measurement.

Studies also comprise some limitations which need to be taken into account when considering the results. One potential limitation of this study is the aspect that the women were matched based on their age, but not concerning their social background. Hence, while 19 % of AN patients had a higher school certificate, this was the case in 57 % of the control women.

A further potential limitation concerns analyses of differences regarding structural changes. All analyses regarding the volume of the brain in total and the various sub-structures (e.g. white and gray matter volume, left and right cortex etc.) were calculated by means of analyses of variance. They could not be analyzed in a multivariate model as the brain regions partially overlap, which affects the validity of the results. In addition to this, data was analyzed in terms of an explorative analysis. This problem also concerns the associations between the volume of the amygdala and other brain regions and co-morbid symptoms of AN, as well as all correlations between the brain activities due to the administration of a taste in general with various psychological aspects (such as anxiety and pleasantness ratings, stress, and co-morbid symptoms).

Another potential limitation of this study is that the fluids had to be administered manually to the participants with syringes because financial support was limited and the corresponding equipment was not available considering the fact that this was the first-ever taste study which was conducted at the Medical University of Graz. As a consequence, the taste stimuli were administered in a blocked design to eliminate consequential measurement errors.

A further potential limitation of this study is that psychological aspects, structural and also functional neurobiological changes in individuals who were currently affected by AN were analyzed. Therefore, it was not possible to differentiate between the primary disorder and the secondary phenomena accompanying the disease, given that malnutrition brings about changes in neuroendocrine function. The cross-sectional design of this study did not allow a determination of whether the findings of this study are traits contributing to the onset of AN or a consequence of malnutrition and weight loss. Studies have shown that with recovery from AN, some brain changes seem to normalize (Fuglset & Frampton, 2011; Mainz et al., 2012; Wagner et al., 2006a), while others persist (Fuglset & Frampton, 2011; Giordano et al., 2001; Golden & Meyer, 2004). Further research would be required as to the question whether the volume loss in certain brain regions and the alterations concerning taste processing and their relation with psychological symptoms persist after a recovery from AN and after associated co-morbidities have disappeared.

4.5 Implications and conclusions

This study showed that patients with AN show certain psychological symptoms and a selective volume loss in certain brain areas. The results also show that the size of certain fields of the brain is associated with certain symptoms and they evidence an altered taste processing together with the fact that taste processing was associated with certain psychological aspects. So far, hardly any information regarding physiological correlates of behavioral disorders has been available due to the inaccessibility of the brain. In addition to this, relatively few studies have analyzed neurobiological changes in people with EDs compared to other psychiatric illnesses (Kaye et al., 2013). Today, imaging methods such as MRI can provide new insights into the structural and functional changes of the brain, which – in addition to environmental interactions – might contribute to the development of AN; however, cause and consequence remain unclear.

EDs, and especially AN, are the most severe psychiatric illness and have the highest mortality rate of all mental disorders (Bulik et al., 2006; Frank et al., 2008; Kaye et al., 2011). Most therapies are still inadequate, considering the high number of relapses and the low recovery rate (Carter et al., 2009; Kaye et al., 2011). Understanding the neurobiological and pathophysiological facts which might contribute to the development and maintenance of the disease might lead to the development of more effective therapies. It remains difficult, however, to directly translate brain findings into therapeutic measures that would improve symptoms in patients, as brain research is still a rather unexplored field and randomized controlled trials to analyze neurobiological changes in the long-run are still lacking.

One consequence of our findings could be the modulation of already existing psychotherapies to better target symptoms which are present in patients with AN such as targeting strategies to react in a more flexible way and to reinterpret food reward, as patients with AN seem to have an overly active satiety signal in response to palatable foods (Oberndorfer et al., 2013). Our results indicate that psychological aspects such as stress, coping, body image issues, achievement orientation, and perfectionism should be targeted in psychotherapy.

In addition to this, another possibility seems to be the use of an fMRI-based neurofeedback training to alter the brain activity in rewarding brain systems such as the insula, which are disturbed in patients with AN – an approach which has already proved to be effective in other disorders (Birbaumer et al., 2013; Caria et al., 2010; Oberndorfer et al., 2013). Additionally, previous studies report that food consumption can be reduced by an alteration concerning thoughts of food (Higgs & Donhoe, 2011; Morewedge et al., 2010).

Overall, the results of this study show that AN is a serious disease which is associated with certain psychological and neuropsychological abnormalities. Considering the general poor prognosis of the disease with a high number of individuals who develop a chronic illness or die (Kaye et al., 2009), this study helps to better understand the pathophysiology of the disease, and, subsequently, can help to find a reasoned approach to its treatment.

5 References

American Psychiatric Association (1994) *Diagnostic and Statistical Manual of Mental Disorders.* Volume 4. Washington DC: American Psychiatric Press.

Andersen, P., Morris, R., Amaral, D., Bliss, T. & O'Keefe, J. (2006) *The Hippocampus Book.* Oxford: University Print.

Anzengruber, D., Klump, K. L., Thornton, L., Brandt, H., Crawford, S., Fichter, M. M., Halmi, K. A., Johnson, C., Kaplan, A. S., LaVia, M., Mitchell, J., Strober, M., Woodside, D. B., Rotondo, A., Berrettini, W. H., Kaye, W. H. & Bulik, C. M. (2006) Smoking in Eating Disorders. *Eating Behaviors,* 7: 291-299.

Aoki, C., Sabaliauskas, N., Chowdhury, T., Min, J. Y., Colacino, A. R. & Laurino, K. (2012) Adolescent female rats exhibiting activity-based anorexia express elevated levels of GABA(A) receptor $\alpha 4$ and δ subunits at the plasma membrane of hippocampal CA1 spines. *Synapse,* 66(5): 391-407.

Apfel, B. A., Ross, J., Hlavin, J., Meyerhoff, D. J., Metzler, T. J., Marmar, C. R., Weiner, M. W., Schuff, N. & Neylan, T. C. (2011) Hippocampal volume differences in Gulf War veterans with current versus lifetime posttraumatic stress disorder symptoms. *Biological Psychiatry,* 69(6): 541-548.

Arnold, M.B. (1960) Emotion and personality. New York: Columbia Press.

Artmann, H., Grau, H., Adelmann, M. & Schleiffer, R. (1985) Reversible and non-reversible enlargement of cerebrospinal fluid spaces in anorexia nervosa. *Neuroradiology,* 27(4): 304-312.

Babaei, A., Kern, M., Antonik, S., Mepani, R., Ward, B. D., Li, S. J., Hyde, J. & Shaker, R. (2010) Enhancing effects of flavorered nutritive stimuli on cortical swallowing network activity. *American Journal of Physiology: Gastrointestinal Liver Physiology,* 299: G422-G429.

Bailer, U. F., Frank, J. K., Henry, S. E., Price, J. C., Meltzer, C. C., Weissfeld, L., Mathis, C. A., Drevets, W. C., Wagner, A., Hoge, J., Ziolko, S. K., McConaha, C. W. & Kaye, W. H. (2006) Altered brain serotonin 5-HT1A receptor binding after recovery from anorexia nervosa measured by positron emission tomography and (carbonyl11 c) WAY-1000635. *Biological Psychiatry,* 59: 291-293.

Bailer, U. & Kaye, W. H. (2010) Serotonin: Imaging Findings in Eating Disorders. In: Adan, R. A. H. & Kaye, W. H. (eds.) *Behavioral Neurobiology of Eating Disorders.* Berlin, Heidelberg: Springer, 59-79.

Barbarich, N. C., Kaye, W. H. & Jimerson, D. (2003) Neurotransmitter and Imaging Studies in Anorexia Nervosa: New Targets for Treatment. *CNS & Neurological Disorders,* 2: 61-72.

Barbarich-Marsteller, N. C., Fornal, C. A., Takase, L. F., Bocarsly, M. E., Arner, C., Walsh, B. T., Hoebel, B. G. & Jacobs, B. L. (2013) Activity-based anorexia is associated with reduced hippocampal cell proliferation in adolescent female rats. *Behavioral Brain Research,* 236(1): 251-257.

Barr, J. L. & Foster, G. L. (2011) Serotonergic neurotransmission in the ventral hippocampus is enhanced by corticosterone and altered by chronic amphetamine treatment. *Neuroscience,* 182: 105-114.

Bartoshuk, L. M., McBurney, D. H. & Pfaffmann, C. (1964) Taste of sodium chloride solutions after adaptation to sodium chloride: implications for the "water taste". *Science,* 143: 967-968.

Beaver, J. D., Lawrence, A. D., van Ditzhuijzen, J., Davis, M. H., Woods, A. & Calder A. J. (2006) Individual differences in reward drive predict neural responses to images of food. *Journal of Neuroscience,* 26: 5160-5166.

Bender, G., Veldhuizen, M. G., Meltzer, J. A., Gitelman, D. R. & Small, D. M. (2009) Neural correlates of evaluative compared with passive tasting. *European Journal of Neuroscience,* 30(2): 327-338.

Bergmann, G., Loewe, B., Zipfel, S., Buchholz, C. & Herzog, W. (2003) Langzeitergebnisse bei Anorexia nervosa nach 21 Jahren in einer prospektiven Follow-up Studie. In: Steinbrenner, B. & Schoenauer-Cejpek, M. (eds.) *Essstörungen. Anorexie-Bulimie-Adipositas. Therapie in Theorie und Praxis.* Vienna, Munich, Bern: Verlag Wilhelm Maudrich, 40-52.

Berthoud, H. R. & Morriso, C. (2008) The brain, appetite, and obesity. *Annual Review of Psychology,* 59: 55-92.

Berthoud, H. R. (2011) Metabolic and hedonic drives in the neural control of appetite: who is the boss? *Current Opinion of Neurobiology,* 21: 888–896.

Birbaumer, N., Ruiz, S., and Sitaram, R. (2013). Learned regulation of brain metabolism. *Trends of Cognitive Science,* 17: 295–302.

Bloks, H., Van Furth, E. F., Callewaert, I. & Hoek, H. W. (2004) Coping strategies and recovery in patients with a severe eating disorder. *Eating Disorders,* 12(2): 157-169.

Bohon, C. & Stice, E. (2010) Reward abnormalities among women with full and subthreshold bulimia nervosa: a functional magnetic resonance imaging study. *International Journal of Eating Disorders,* 44: 585-595.

Bosanac, P., Kurlender, S., Stojanovska, L., Hallman, K., Norman, T., McGrath, C., Burrows, G., Wesnes, K., Manktelow, T. & Olver, J. (2007) Neuropsychological study of underweight and "weight-recovered" anorexia nervosa compared with bulimia nervosa and normal controls. *International Journal of Eating Disorders,* 40: 613-621.

Bossert, S., Laessle, R. G., Meiller, C., Junker, M., Ellgring, H. & Pirke, K. M. (1991) Visual palatability of food in patients with eating disorders and dieting women. *Behavioral Research and Therapy,* 29: 337-341.

Brooks, S. J., Barker, G. J., O'Daly, O. G., Brammer, M., Williams, S. C., Benedict, C., Schioeth, H. B., Treasure, J. & Campbell, I. C. (2011) Restraint of appetite and reduced regional blood volumes in anorexia nervosa: a voxel-based morphometric study. *BMC Psychiatry,* 11: 179.

Brown, V. M. & Morey, R. A. (2012) Neural systems for cognitive and emotional processing in posttraumatic stress disorder. *Frontiers in Psychology,* 3: 449.

Brytek, A. (2006) Self-esteem, strategies of coping and feeling of anger in French patients with Anorexia Nervosa. *Psychiatroa Polska,* 40(4): 743-750.

Brytek-Matera, A. & Schlitz, L. (2013) Comparative structural study of the configuration of coping strategies among female patients with eating disordes and a non-clinical control group. *Psychiatria Danubina,* 25(4): 359-365.

Bulik, C. M., Sullivan, P. F. & Joyce, P. R. (1999) Temperament, character and suicide attempts in anorexia nervosa, bulimia nervosa and major depression. *Acta Psychiatrica Scandinavica,* 100(1): 27-32.

Bulik, C. M., Sullivan, P. F., Wade, T. D. & Kendler, K. S. (2000) Twin studies of eating disorders: a review. *International Journal of Eating Disorders,* 27: 1-20.

Bulik, C. M., Sullivan, P. F., Tozzi, F., Furberg, H., Lichtenstein, P. & Pedersen, N. L. (2006) Prevalence, Heritability, and Prospective Risk Factors for Anorexia Nervosa. *Archives of General Psychiatry,* 63: 305-312.

Bulik, C. M., Thornton, L., Pinheiro, K., Klump, K. L., Brandt, H., Crawford, S., Fichter, M. M., Halmi, K. A., Johnson, C., Kaplan, A. S., Mitchell, J., Nutzinger, D., Strober, M., Treasure, J., Woodside, D. B., Berrettini, W. H. & Kaye, W. H. (2008) Suicide attempts in anorexia nervosa. *Journal of Psychsomatic Medicine,* 70(3): 378-383.

Burger, K. S. & Stice, E. (2011) Relation of dietary restraint scores to activation of reward-related brain regions in response to food intake, anticipated intake, and food pictures. *NeuroImage,* 55: 233-239.

Burkert, N. T., Koschutnig, K., Ebner, F. & Freidl, W. (2015) Structural hippocampal alterations, perceived stress, and coping deficiencies in patients with anorexia nervosa. *International Journal of Eating Disorders,* epub ahead of print. DOI: 10.1002/eat.22397.

Cameron, J. D., Goldfield, G. S., Cyr, M. J. & Doucet, E. (2008) The effects of prolonged caloric restriction leading to weight-loss on food hedonics and reinforcement. *Physiology & Behavior*, 94: 474-480.

Campbell, S. & Macqueen, G. (2004) The role of the hippocampus in the pathophysiology of major depression. *Journal of Psychiatry & Neuroscience*, 29(6): 417-426.

Caria, A., Sitaram, R., Veit, R., Begliomini, C. & Birbaumer, N. (2010) Volitional control of anterior insula activity modulates the response to aversive stimuli. A real-time functional magnetic resonance imaging study. *Biological Psychiatry*, 68: 425-432.

Carter, J. C., McFarlane, T. L., Bewell, C., Olmsted, M. P., Woodside, D. B., Kaplan, A. S. & Crosby, R. D. (2009) Maintenance treatment for anorexia nervosa: a comparison of cognitive behavior therapy and treatment as usual. *International Journal of Eating Disorders, 42*(3): 202-207.

Castro, J., Deulofeu, R., Gila, A., Puig, J. & Toro, J. (2004) Persistence of nutritional deficiencies after short term recovery in adolescents with anorexia nervosa. *International Journal of Eating Disorders*, 35: 169-178.

Castro-Fornieles, J., Bargalló, N., Lazaro, L., Andres, S., Falcon, C., Plana, M. T. & Junqué, C. (2007) Adolescent anorexia nervosa: A cross-sectional and follow-up frontal grey matter disturbances detected with proton magnetic resonance spectroscopy. *Journal of Psychiatric Researcvh*, 41: 952-958.

Castro-Fornieles, J., Caldú, X., Andrés-Perpiná, S., Lázaro, L., Bargalló, N., Falcón, C., Plana, M. T. & Junqué, C. (2010) A cross-sectional and follow-up functional MRI study with a working memory task in adolescent anorexia nervosa. *Neuropsychologia*, 48: 4111-4116.

Cerf-Ducastel, B. & Murphy, C. (2004) Validation of a stimulus protocol suited to the investigation of odor-taste interactions with fMRI. *Physiological Behaviour*, 81(3): 389-396.

Channon, S., Hemsley, D. & de Silva, P. (1988) Selective processing of food words in anorexia nervosa. *British Journal of Clinical Psychology*, 27: 259-260.

Childress, J. E., McDowell, E. J., Dalai, V. V., Bogale, S. R., Ramamurthy, C., Jawaid, A., Kunik, M. E., Qureshi, S. U. & Schulz, P. E. (2013) Hippocampal volumes in patients with chronic combat-related posttraumatic stress disorder: a systematic review. *The Journal of Neuropsychiatry and Clinical Neurosciences,* 25(1): 12-25.

Chowdhury, T. G., Wable, G. S., Sabaliauskas, N. A. & Aoki, C. (2013) Adolescent female C57BL/6 mice with vulnerability to activity-based anorexia exhibit weak inhibitory input onto hippocampal CA1 pyramidal cells. *Neuroscience,* 241C: 250-267.

Chui, H. T., Christensen, B. K., Zipursky, R. B., Richards, B. A., Hanratty, M. K., Kabani, N. J., Mikulis, D. J. & Katzman, D. K. (2008) Cognitive function and brain structure in females withz a history of adolescent-onset anorexia nervosa. *Pediatrics,* 122(2):e426-e437.

Cohen, J. (1992) A power primer. *Psychol Bulletin,* 112: 155-159.

Comings, D. E. & Blum, K. (2000) Reward deficiency syndrome: genetic aspects of behavioral disorders. *Progress in brain research,* 126: 325-341.

Connan, F., Murphy, F., Connor, S. E., Rich, P., Murphy, T., Bara-Carill, N., Landau, S., Krljes, S., Ng, V., Williams, S., Morris, R. G., Campbell, I. C. & Treasure, J. (2006) Hippocampal volume and cognitive function in Anorexia Nervosa. *Psychiatry Research,* 146(2): 117-125.

Cooley, E. & Toray, T. (2001) Body image and personality predictors of eating disorder symptoms during the college years. *International Journal of Eating Disorders,* 30: 28-36.

Corcos, M., Taieb, O., Benoit-Lamy, S., Paterniti, S., Jeammet, P. & Flament, M. F. (2002) Suicide attempts in women with bulimia nervosa: frequency and characteristics. *Acta Psychiatria Scandinavia,* 106: 381-386.

Cowdry, F. A., Park, R. J., Harmer, C. J. & McCabe, C. (2011) Increased neural processing of rewarding and aversive food stimuli in recovered anorexia nervosa. *Biological Psychiatry,* 70(8): 736-743.

Craig, A. D. (2009) How do you feel – now? The anterior insula and human awareness. *Nature Reviews,* 10: 59-70.

Crow, S. J., Peterson, C. B., Swanson, S. A., Raymond, N. C., Specker, S., Eckert, E. D. & Mitchell, J. E. (2009) Increased mortality in bulimia nervosa and other eating disorders. *American Journal of Psychiatry,* 166: 1342-1346.

Dannlowski, U., Stuhrmann, A., Beutelmann, V., Zwanzger, P., Lenzen, T., Grotegerd, D., Domschke, K., Hohoff, C., Ohrmann, P., Bauer, J., Lindner, C., Postert, C., Konrad, C., Arolt, V., Heindel, W., Suslow, T. & Kugel, H. (2012) Limbic scars: long-term consequences of childhood maltreatment revealed by functional and structural magnetic resonance imaging. *Biological Psychiatry,* 71(4): 286-293.

Davis, R., Freeman, R. J. & Garner, D. M. (1988) A naturalistic investigation of eating behavior in bulimia nervosa. *Journal of Consulting and Clinical Psychology,* 56: 273-279.

Davis, C., Strachan, S. & Berkson, M. (2004) Sensitivity to reward: Implications for overeating and obesity. *Appetite,* 42: 131-138.

Dawe, S. & Loxton, N. J. (2004) The role of impulsivity in the development of substance use and eating disorders. *Neuroscience & Biobehavioral Reviews,* 28: 343-351.

De Araujo, I. E. T., Kringelbach, M. L., Rolls, E. T. & Hobden, P. (2003a) The representation of umami taste in the human brain. *Journal of Neurophysiology,* 90: 313-319.

De Araujo, I. E. T., Kringelbach, M. L., Rolls, E. T. & McGlone, F. (2003b) Human cortical responses to water in the mouth, and the effects of thirst. *Journal of Neurophysiology,* 90: 1865-1876.

De Araujo, I. E. T. & Rolls, E. T. (2004) The representation in the human brain of food texture and oral fat. *Journal of Neurosciences,* 24: 3086-3093.

Delevenee, V., Lotstra, F., Goldman, S., Biver, F., De Maertelaer, V., Appelboom-Fondu, J., Schoutens, A., Bidaut, L. M., Luxen, A. & Mednelwicz, J. (1995) Brain hypometabolism of glucose in anorexia nervosa: a PET scan study. *Biological Psychiatry,* 37(3): 161-169.

Delevenne, V., Goldman, S., De Maertelaer, V., Simon, Y., Luxen, A. & Lotstra, F. (1996) Brain hypometabolism of glucose in anorexia nervosa: normalizarion after weight gain? *Biological Psychiatry,* 40(8): 761-768.

Delevenne, V., Goldman, S., De Maertelaer, V., Wikler, D., Damhaut, P. & Lotstra, F. (1997) Brain glucose metabolism in anorexia nervosa and affective disorders: influence of weight loss or depressive symptomatology? *Psychiatry Research,* 74(2): 83-92.

Delparigi, A., Chen, K., Salbe, A. D., Reiman, E. M. & Tataranni, P. A. (2005) Sensory experience of food and obesity: a positron emission tomography study of the brain regions affected by tasting a liquid meal after a prolonged fast. *NeuroImage,* 24: 436-443.

Desmond, J. E. & Glover, G. (2002) Estimating sample size in functional MRI (fMRI) neuroimaging studies: Statistical power analyses. *Journal of Neuroscience Methods,* 118: 115-128.

De Vos, J., Houtzager, L., Katsaragaki, G., van de Berg, E., Cuijpers, P. & Dekker, J. (2014) Meta analysis on the efficacy of pharmacotherapy versus placebo on anorexia nervosa. *Journal of Eating Disorders,* 2(1): 27.

De Zwaan, M. (2001) Binge eating and obesity. *International Journal of Obesity,* 25: 51-55.

De Zwaan, M,, Nutzinger, D. O. & Schoenbeck, G. (1992) Binge eating in overweight women. *Comprehensive Psychiatry,* 33: 256-261.

Diamanti, A., Basso, M. S., Castro, M., Bianco, G., Ciacco, E., Calce, A., Caramadre, A. M., Noto, C. & Gambarara, M. (2008) Clinical Efficiacy and Safety of Parental Nutrition in Adolescent Girls with Anorexia Nervosa. *Journal of Adolescence Health,* 42: 111-118.

Ding, A. Y., Li, Q., Zhou, I. Y., Ma, S. J., Tong, G., McAlonan, G. M. & Wu, E. X. (2013) MR diffusion tensor imaging detects rapid microstructural changes in amygdala and hippocampus following fear conditioning in mice. *PLoS One,* 8(1): e51704.

Duvvuri, V. & Kaye, W. H. (2009) Anorexia Nervosa. *FOCUS,* 7(4): 455-462.

Eberly, M. C. (2005) Understand Self-Injurious Behavior in Eating Disorders. *The Remuda Review: The Christian Journal of Eating Disorders,* 4: 26-30.

Ehrig, C. (2003) Stationäre Verhaltenstherapie von Essstörungen am Beispiel der Anorexia nervosa. In Steinbrenner, B. & Schoenauer-Cejpek, M. (eds.) *Essstörungen. Anorexie-Bulimie-Adipositas. Therapie in Theorie und Praxis.* Vienna, Munich, Bern: Verlag Wilhelm Maudrich, 53-65.

Ellison, Z., Foong, J., Howard, R., Bullmore, E., Williams, S. & Treasure, J. (1998) Functional anatomy of calorie fear in anorexia nervosa. *The Lancet,* 352: 1192.

Engel, G. L. (1977) The need for a new medical model: A challenge for biomedicine. *Science,* 196: 129-136.

Engel, G. L. (1980) The clinical application of the biopsychosocial model. *American Journal of Psychiatry,* 137(5): 535-544.

Erdmann, G. & Janke, W. (2008) *SVF. Stressverarbeitungsfragebogen. Stress, Stressverarbeitung und ihre Erfassung durch ein mehrdimensionales Testsystem.* Goettingen: Hogrefe.

Espeset, E. M., Gulliksen, K. S., Nordbo, R. H., Skarderud, F. & Holte, A. (2012) The link between negative emotions and eating disorder behavior in patients with anorexia nervosa. *European Journal of Eating Disorder Reviews* 20(6): 451-460.

Espinoza, P., Penelo, E. & Raich, R. M. (2009) Disordered eating behaviors and body image in a longitudinal pilot study of adolescent girls: What happens 2 years later? *Body Image,* 7(1): 70-73.

Fahrenberg, J., Hampel, R. & Selg, H. (1984) *Das Freiburger Persönlichkeitsinventar FPI. Revidierte Fassung FPI-R und teilweise geänderte Fassung FPI-A1. Handanweisung.* 4. revidierte Auflage. Goettingen: Hogrefe.

Fairburn, C. G. & Harrison, P. J. (2003) Eating disorders. *The Lancet,* 361: 407-416.

Favaro, A., Tenconi, E. & Santonastaso, P. (2006) Perinatal factors and the risk of developing anorexia nervosa and bulimia nervosa. *Archives of Genral Psychiatry,* 63: 82-88.

Favaro, A., Santonastaso, P., Manara, R., Bosello, R., Bommarito, G., Tenconi, E. & Di Salle, F. (2012) Disruption of visuospatial and somatosensory functional connectivity in anorexia nervosa. *Biological Psychiatry,* 72(10): 864-870.

Feldman, L. A., Shapiro, M. L. & Nalbantoglu, J. (2010) A novel, rapidly acquired and persistent spatial memory task that induces immediate early gene expression. *Behavioral Brain Function,* 6: 35.

Figlewicz, D. P. & Benoit, S. C. (2009) Insulin, leptin, and food reward: update 2008. *American Journal of Physiology – Regulartory, Integrative and Comparative Physiology,* 296: R9-R19.

Fischl, B., Salat, D. H., Busa, E., Albert, M., Dieterich, M., Haselgrove, C., van der Kouwe, A., Killiany, R., Kennedy, D., Klaveness, S., Montillo, A., Makris, N., Rosen, B. & Dale, A. M. (2002) Whole brain segmentation: automated labeling of neuroanatomical structures in the human brain. *Neuron,* 33: 341-355.

Fladung, A. K., Grön, G., Grammer, K., Hermberger, B., Schilly, E., Grasteit, S., Wolf, R. C., Walter, H. & von Wietersheim, J. (2010) A neural signature of anorexia nervosa in the ventral striatal reward system. *American Journal of Psychiatry,* 167: 206-212.

Foulon, C., Guelfi, J. D., Kipman, A., Adès, J., Romo, L., Houdeyer, K., Marquez, S., Mouren, M. C., Rouillon, F. & Gorwood, P. (2007) Switching to the binging/purging subtype of anorexia nervosa is frequently associated with suicidal attempts. *European Journal of Psychiatry,* 22: 513-519.

Fowler, L., Blackwell, A., Jaffa, A., Palmer, R., Robbins, T. W., Sahakian, B. J. & Dowson, J. H. (2006) Profile of neurocognitive impairments associated with female in-patients with anorexia nervosa. *Psychological Medicine,* 36: 517-527.

Frampton, I., Watkins, B., Gordon, I. & Lask, B. (2011) Do abnormalities in regional blood flow in anorexia nervosa resolve after weight restoration? *European Eating Disorder Reviews,* 19(1): 55-58.

Frank, G. K., Kaye, W. H., Carter, C. S., Brooks, S., May, C., Fissell, K. & Stenger, V. A. (2003) The evaluation of brain activity in response to taste stimuli – A pilot study and method for central taste activation as assessed by event related fMRI. *Journal of Neuroscience Methods,* 131: 99-105.

Frank, G. K., Bailer, U. F., Henry, S., Wagner, A. & Kaye, W. H. (2004) Neuroimaging studies in Eating Disorders. *CNS Spectrums,* 9(7): 539-548.

Frank, G. K., Wagner, A., Achenbach, S., McConaha, C., Skovira, K., Aizenstein, H., Carter, C. S. & Kaye, W. H. (2006) Altered brain activity in women recovered from bulimic-type eating disorders after a glucose challenge: A pilot study. *International Journal of Eating Disorders,* 39: 76-79.

Frank, G. K., Oberndorfer, T. A., Simmons, A. N., Paulus, M. P., Fudge, J. L., Yang, T. T. & Kaye, W. H. (2008) Sucrose activates human taste pathways differently from artificial sweetener. *NeuroImage,* 39: 1559-1569.

Frank, G. K., Reynolds, J. R., Shott, M. E., Jappe, L., Yang, T. T., Tregellas, J. R. & O'Reilly, R. C. (2012) Anorexia Nervosa and Obesity are associated with Opposite Brain Reward Response. *Neuropsychopharmacology,* 37(9): 2031-2046.

Frank, G. K., Shott, M. E., Hagman, J. O. & Mittal, V. A. (2013) Alterations in Brain Structures related to taste reward circuitry in ill and recovered Anorexia Nervosa and in Bulimia Nervosa. *American Journal of Psychiatry,* 170: 1152-1160.

Frank, S., Laharnar, N., Kullmann, S., Veit, R., Canova, C., Hegner, Y. L., Fritsche, A. & Preissl, H. (2010) Processing of food pictures: Influence of hunger, gender and caloric content. *Brain Research,* 1350: 159-166.

Frank, S., Kullmann, S. & Veit, R. (2013) Food related processes in the insula cortex. *Frontiers in Human Neuroscience,* 7: 499.

Franke, G. H. (1995) Die *Symptom-Checkliste von Derogatis. Deutsche Version (SCL-90-R). Manual.* Goettingen: Beltz Test GmbH.

Franko, D. L. & Keel, P. K. (2006) Suicidality in eating disorders: occurrence, correlates, and clinical implications. *Clinical Psycholpgy Reviews,* 26(6): 769-782.

Frey, S. & Petrides, M. (1999) Re-examination of the human taste region: a positron emission tomography study. *European Journal of Neuroscience,* 11: 2985-2988.

Friederich, H. C., Brooks, S., Uher, R., Campbell, I. C., Giampietro, V., Brammer, M., Williams, S. C., Herzog, W. & Treasure, J. (2010) Neural correlates of body dissatisfaction in anorexia nervosa. *Neuropsychologia,* 48: 2878-2885.

Friedrich, H. C., Walther, S., Bendszus, M., Biller, A., Thomann, P., Zeigermann, S., Katus, T., Brunner, R., Zastrow, A. & Herzog, W. (2012) Grey matter abnormalities within corticolimbic-striatal circuits in acute and weight-restored anorexia nervosa patients. *NeuroImage,* 59(2): 1106-1113.

Friedrich, H. C., Wu, M., Simon, J. J. & Herzog, W. (2013) Neurocircuit Function in Eating Disorders. *International Journal of Eating Disorders,* 46(5): 425-432.

Frieling, H., Fischer, J., Wilhelm, J., Engelhorn, T., Bleich, S., Hillemacher, T., Doerfler, A., Kornhuber, J., de Zwaan, M. & Peschel, T. (2012) Microstructural abnormalities of the posterior thalamic radiation and the mediodorsal thalamic nuclei in females with anorexia nervosa – a voxel based diffusion tensor imaging (DTI) study. *Journal of Psychiatric Research,* 46(9): 1237-1242.

Fuglset, T. S. & Frampton, I. (2011) Neuroimaging. In Lask, B. & Frampton, I. (eds.) *Eating Disorders and the Brain.* Chichester: Wiley-Blackwell, 56-105.

Fuster, J. M. (1993) Frontal lobes. *Current Opinion in Neurobiology,* 3: 160-165.

Gaudio, S. & Quattrocchi, C. C. (2012) Neural basis of a multidimensional model of body image distortion in anorexia nervosa. *Neuroscience & Biobehavarioral Reviews,* 36(8): 1839-1847.

Gaudio, S., Nocchi, F., Franchin, T., Genovese, E., Cannatà, V., Longo, D. & Fariello, G. (2011) Gray matter desrease distribution in the early stages of anorexia nervosa restrictive type in adolescents. *Psychiatry Research,* 191(1): 24-30.

Geliebter, A., Ladell, T., Logan, M., Schneider, T., Sharafi, M. & Hirsch, J. (2006) Responsivity to food stimuli in obese and lean binge eaters using functional MRI. *Appetite,* 46: 31-35.

George, H. R., Cornelissen, P. L., Hancock, P. J., Kiviniemi, V. V. & Tovée, M. J. (2011) Differences in eye-movement patterns between anorexic and control observers when judging body size and attractiveness. *British Journal of Psychology,* 102(3): 340-354.

Giordano, G. D., Renzetti, P., Parodi, R. C., Foppiani, L., Zandrino, F., Giordano, G. & Sardanelli, F. (2001) Volume measurement with magnetic resonance imaging of hippocampus-amygdala formation in patients with Anorexia Nervosa. *Journal of Endocrinological Investigation,* 24(7): 510-514.

Golden, N. H. & Meyer, W. (2004) Nutritional rehabilitation of anorexia nervosa. Goals and dangers. *International Journal of Adolescent Medicine and Health,* 16(2): 131-144.

Goldstone, A. P., Prechtl de Hernandez, C. G., Beaver, J. D., Muhammed, K., Croese, C., Bell, G., Durighel, G., Hughes, E., Waldman, A. D., Frost, G. & Bell, J. D. (2009) Fasting biases brain reward systems toward high-caloric foods. *European Journal of Neuroscience,* 30: 1625-1635.

Gordon, C. M., Dougherty, D. D., Rauch, S. L., Emans, S. J., Grace, E., Lamm, R., Alpert, N. M., Majzoub, J. A. & Fischman, A. J. (2000) Neuroanatomy of human appetitive function: a positron emission tomography investigation. *International Journal of Eating Disorders,* 27: 163-171.

Gottfried, J. A., O'Doherty, J. & Dolan, R. J. (2002) Appetitive and aversive olfactory learning in humans studied using event-related functional magnetic resonance imaging. *Journal of Neuroscience,* 22: 10829-10837.

Gottfried, J. A., O'Doherty, J. & Dolan, R. J. (2003) Encoding predictive reward value in human amygdala and orbitofrontal cortex. *Science,* 301: 1104-1107.

Grill, H. J., Skibicka, K. P. & Hayes, M. R. (2007) Imaging obesity: fMRI, food reward, and feeding. *Cell Metabolism,* 6: 423-425.

Guardia, D., Cottencon, O., Thomas, P., Dodin, V. & Luyat, M. (2012) Spatial orientation constancy is impaired in anorexia nervosa. *Psychiatry Research,* 195(1-2): 56-59.

Hall, T., Galletly, C., Clark, C. R., Veltmeyer, M., Metzger, L. J., Gilbertson, M. W., Orr, S. P., Pitman, R. K. & McFarlane, A. (2012) The relationship between Hippocampal asymmetry and working memory processing in combat-related PTSD - a monozygotic twin study. *Biology of Mood & Anxiety Disorders,* 2(1): 21.

Hatch, A., Madden, S., Kohn, M. R., Clarke, S., Touyz, S., Gordon, E. & Williams, L. M. (2010) In first presentation adolescent anorexia nervosa, do cognitive markers of underweight status change with weight gain following a refeeding intervention? *International Journal of Eating Disorders,* 43: 295-306.

Hautzinger, M., Bailer, M., Worall, H. & Keller, F. (1995) *BDI. Beck-Depressions-Inventar. Testhandbuch.* 2. ueberarbeitete Auflage. Bern: Verlag Hans Huber.

Hay, P. J. & Sachdev, P. (2011) Brain dysfunction in anorexia nervosa: cause or consequence of under-nutrition? *Current Opinion in Psychiatry,* 24: 251-256.

Heechul, J., Hussaini, S. M. Q., Rigby, M. J. & Jang, M. H. (2012) Functional role of adult hippocampal neurogenesis as a therapeutic strategy for mental disorders. *Neural Plasticy,* 2012:854285.

Henning-Fast, K., Werner, N. S., Lermer, R., Latscha, K., Meister, F., Reiser, M., Engel, R. R. & Meindl, T. (2009) After facing traumatic stress: brain activation, cognition and stress coping in policemen. *Journal of Psychiatric Research,* 43(14): 1146-1155.

Herzog, D. B., Greenwood, D. N., Dorer, D. J., Flores, A. T., Ekeblad, E. R., Richards, A., Blais, M. A. & Keller, M. B. (2000) Mortality in eating disorders: A descriptive study. *International Journal of Eating Disorders,* 28(1): 20-26.

Higgs, S. & Donohoe, J. E. (2011). Focusing on food during lunch enhances lunch memory and decreases later snack intake. *Appetite, 57*: 202–206.

Hinton, E. C., Parkinson, J. A., Holland, A. J., Arana, F. S., Roberts, A. C. & Owen, A. M. (2004) Neural contributions to the motivational control of appetite in humans. *European Journal of Neuroscience,* 20: 1411-1418.

Hoang, U., Goldacre, M. & James, A. (2014) Mortality following hospital discharge with a diagnosis of eating disorder: National record linkage study, England, 2001-2009. *International Journal of Eating Disorders,* 47(5): 507-515.

Holliday, J., Tchanturia, K., Landau, S., Collier, D. & Treasure, J. (2005) Is impaired set-shifting an endophenotype of anorexia nervosa? *American Journal of Psychiatry,* 162: 2269-2275.

Holm-Denoma, J. M., Witte, T. K., Gordon, K. G., Herzog, D. B., Franko, D. L., Fichter, M., Quadflieg, N. & Joiner, T. E. (2008) Death by suicide among individuals with anorexia as arbiters between explanations of the anorexia-suicide link. *Journal of Affective Disorders,* 107(1-3): 231-236.

Holsen, L. M., Zarcone, J. R., Thompson, T. I., Brooks, W. M., Anderson, M. F., Ahluwalia, J. S., Nollen, N. L. & Savage, C. R. (2005) Neural mechanisms underlying food motivation in children and adolescents. *NeuroImage,* 27: 669-676.

Hudson, J. I., Hiripi, E., Pope, H. G. & Kessler, R. C. (2007) The prevalence and correlates of eating disorders in the National Co-morbidity Survey Replication. *Biological Psychiatry,* 61: 348-358.

Jappe, L. M., Cao, L., Crosby, R. D., Crow, S. J., Peterson, C. B., Le Grange, D., Engel, S. G. & Wonderlich, S. A. (2014) Stress and eating disorder behavior in Anorexia Nervosa as a function of menstrual cycle status. *International Journal of Eating Disorders,* 47(2): 181-188.

Joels, M. (2008) Functional actions of corticosteroids in the hippocampus. *European Journal of Pharmacology,* 583(2-3): 312–321.

Johnson, J. G., Cohen, P., Kasen, S. & Brook, J. S. (2002) Childhood adversities associated with risk for eating disorders or weight problems during adolescence or early adulthood. *American Journal of Psychiatry,* 159: 394-400.

Kaess, M., Brunner, R. & Plener, P. L. (2011) *Validierung der deutschsprachigen Fassung des Instrumentes zur Erhebung selbstverletzenden Verhaltens: Das Self-Injurious Thoughts and Behaviours Interview (SITBI-G).* Available at: https://www.klinikum.uni-heidelberg.de/SITBI.127031.0.html?&L=3 [accessed on December 04, 2014].

Kaplan, A. S. (2005) From Genes To Treatment Response: New Research into the Psychobiology of Anorexia Nervosa. *International Journal of Eating Disorders,* 37: S89-S89.

Karhunen, L. J., Vanninen, E. J., Kuikka, J. T., Lappalainen, R. I., Tiihonen, J. & Uusitupa, M. J. I. (2000) Regional cerebral blood flow during exposure to food in obese binge eating women. *Psychiatry Research: Neuroimaging,* 99: 29-42.

Katzman, D. K., Lambe, E. K., Mikulis, D. J., Ridgley, J. N., Goldbloom, S. D. & Zipursky, R. B. (1996) Cerebral grey matter and white matter volume deficits in adolescent girls with anorexia nervosa. *Journal of Pediatrics,* 129: 794-803.

Katzman, D. K., Zipursky, R. B., Lambe, E. K. & Mikulis, D. J. (1997) A longitudinal magnetic resonance imaging study of brain changes in adolescents with anorexia nervosa. *Archives of Pediatric and Adolescent Medicine,* 51: 793-797.

Katzman, D. K. & Golden, N. H. (2009) Anorexia Nervosa and Bulimia Nervosa, Chapter 33. In Neinstein, L. S., Gordon, C. M., Katzman, D. K., Rosen, D. S. & Woods, E. R. (eds.) *Handbook of Adolescent Health Care.* Philadelphia: Lippincott, Williams and Wilkins, 321-332.

Kaye, W. H., Greeno, C. G., Moss, H., Fernstrom, J., Fernstrom, M., Lilenfeld, L. R., Weltzin, T. E. & Mann, J. J. (1998) Alterations in Serotonin Activity and Psychiatric Symptoms After Recovery From Bulimia Nervosa. *Archives of General Psychiatry,* 55: 927-935.

Kaye, W. H., Frank, G. K., Meltzer, C. C., Price, J. C., McConaha, C. W., Crossan, P. J., Klump, K. L. & Rhodes, L. (2001) Altered serotonin 2A receptor activity in women who have recovered from bulimia nervosa. *American Journal of Psychiatry*, 158: 1152-1155.

Kaye, W. H., Bailer, U. F., Frank, G. K., Wagner, A. & Henry, S. E. (2005) Brain imaging of serotonin after recovery from anorexia and bulimia nervosa. *Physiology & Behavior*, 86: 15-17.

Kaye, W. H., Fudge, J. L. & Paulus, M. (2009) New insights into symptoms and neurocircuit function of anorexia nervosa. *Nature Reviews*, 10: 573-584.

Kaye, W. H., Wagner, A., Fudge, J. L. & Paulus, M. (2011) Neurocircuity of eating disorders. *Current Topics in Behavioral Neurosciences*, 6: 37-57.

Kaye, W. H., Wierenga, C. E., Bailer, U. F., Simmons, A. N. & Bischoff-Grethe, A. (2013) Nothing tastes as good as skinny feels: the neurobiology of Anorexia Nervosa. *Trends in Neurosciences*, 36(2): 110-120.

Kaye, W. H., Wierenga, C. E., Knatz, S., Liang, J., Boutelle, K., Hill, L. & Eisler, I. (2015) Temperament-based treatment for Anorexia Nervosa. *European Eating Disorders Review*, 23(1): 12-18.

Kazlouski, D., Rollin, M. D., Tregllas, J., Shott, M. E., Jappe, L. M., Hagman, J. O., Pryor, T., Yang, T. T. & Frank, G. K. W. (2011) Altered fimbria-fornix white matter integrity in Anorexia Nervosa predicts harm avoidance. *Psychiatry Research*, 192(2): 109-116.

Keel, P. K., Dorer, D. J., Eddy, K. T., Franko, D., Charatan, D. L. & Herzog, D. B. (2003) Predictors of mortality in eating disorders. *Archives of General Psychiatry*, 60: 179-183.

Kendler, K. S. & Gardner, C.O. (1998) Twin studies of adult psychiatric and substance dependence disorders: are they biased by differences in the environmental experiences of monozygotic and dizygotic twins in childhood and adolescence? *Psychological Medicine*, 28: 625-633.

Kerem, N. C. & Katzman, D. K. (2003) Brain structure and function in adolescents with anorexia nervosa. *Adolescent Medicine (Philadelphia),* 14(1): 109-118.

Keys, A., Brozek, J., Henschel, A., Mickelsen, O. & Taylor, H. L. (1950) *The biology of human starvation* (2 vols). Oxford: University of Minnesota Press.

Killen, J. D., Taylor, C. B., Hayward, C., Haydel, K. F., Wilson, D. M., Hammer, L., Kraemer, H., Blair-Greiner, A. & Strachowski, D. (1996) Weight concerns influence the development of eating disorders: A 4-year prospective study. *Journal of Consulting and Clinical Psychology,* 64: 936-940.

Killgore, W. D., Young, A. D., Femia, L. A., Bogorodzki, P., Rogowska, J. & Todd, D. A. (2003) Cortical and limbic activation during viewing of high- versus low-calorie foods. *NeuroImage,* 19: 1381-1394.

Killgore, W. D. & Yurgelun-Todd, D. A. (2006) Affect modulates appetite-related brain activity to images of food. *International Journal of Eating Disorders,* 39: 357-363.

Kingston, K., Szmukler, G., Andrews, D., Tress, B. & Desmond, P. (1996) Neuropsychological and structural brain changes in anorexia nervosa before and after refeeding. *Psychological Medicine,* 26: 15-28.

Kitamura, A., Tsurugizawa, T., Uematsu, A., Torii, K. & Uneyama, H. (2012) New therapeutic strategy for amino acid medicine: effects of dietary glutamate on gut and brain function. *Journal of Pharmacological Sciences,* 118(2): 138-144.

Klepsch, R., Zaworka, W., Hand, I., Luenenschloß, K. & Jauering, G. (1993) *HZI-K. Hamburger Zwangsinventar – Kurzform.* Goettingen: Beltz Test GmbH.

Klump, K. L., Bulik, C. M., Pollice, C., Halmi, K. A., Fichter, M. M., Berrettini, W. H., Devlin, B., Strober, M., Kaplan, A., Woodside, D. B., Treasure, J., Shabbout, M. M. S., Lilenfeld, L. R. R., Plotnicov, K. H. & Kaye, W. H. (2000) Temperament and Character in Women with Anorexia Nervosa. *The Journal of Nervous and Mental Disease,* 188(9): 559-567.

Kopelman, P. G. (2000) Obesity as a medical problem. *Nature,* 404: 635-643.

Kornreich, L., Shapira, A., Horev, G., Danziger, Y., Tyano, S. & Mimouni, M. (1991) Evaluation of the brain in patients with anorexia nervosa. *American Journal of Neuroradiology,* 6: 1213-1216.

Kringelbach, M. L., O'Doherty, J., Rolls, E. T. & Andrews, C. (2003) Activation of the human orbitofrontal cortex to a liquid food stimulus is correlated with its subjective pleasantness. *Cerebral Cortex,* 13: 1064-1071.

LaBar, K. S., Gitelman, D. R., Parrish, T. B., Kim, Y. H., Nobre, A. C. & Mesulam, M. M. (2001) Hunger selectively modulatres corticolimbic activation to food stimuli in humans. *Behavioral Neuroscience,* 115: 493-500.

Lahousen, T., Hofmann, P., Knoflach-Reichart, C. & Bonelli, R. (2003) Psychopharmakologische Behandlung von Essstörungen. In Steinbrenner, B. & Schoenauer-Cejpek, M. (eds.) *Essstörungen. Anorexie-Bulimie-Adipositas. Therapie in Theorie und Praxis.* Vienna, Munich, Bern: Verlag Wilhelm Maudrich, 9-19.

Lauer, C. J., Gorzewski, B., Gerlinghoff, M., Backmund, H. & Zih, l. J. (1999) Neuropsychological assessments before and after treatment in patients with anorexia nervosa and bulimia nervosa. *Journal of Psychiatric Research,* 33: 129-138.

Laux, L., Glanzmann, P., Schaffner, P. & Spielberger, C. D. (1981) *STAI. Das State-Trait-Angstinventar. Theoretische Grundlagen und Handanweisung.* Weinheim: Beltz Test GmbH.

Lavenex, P. & Lavenex, B. P. (2013) Building hippocampal circuits to learn and remember: insights into the development of human memory. *Behavioral Brain Research,* 254: 8-21.

Lazarus, R. S. (1991) Emotion and adaption. New York: Oxford University Press.

Lemberg, R. (Ed.) (1999) *Eating disorders – a reference sourcebook.* Phoenix: Oryx Press.

Lilenfeld, L. R., Kaye, W. H., Greeno, C. G., Merikangas, K. R., Plotnicov, K., Pollice, C., Rao, R., Strober, M., Bulik, C. M. & Nagy, L. (1998) A Controlled Family Study of Anorexia Nervosa and Bulimia Nervosa. *Archives of General Psychiatry,* 55: 603-610.

Lindauer, R. J., Vlieger, E. J., Jalink, M., Olff, M., Carlier, I. V., Majoie, C. B., Den Heeten, G. J. & Gersons, B. P. (2005) Effects of psychotherapy on hippocampal volume in out-patients with post-traumatic stress disorder: a MRI investigation. *Psychological Medicine,* 35(10): 1421-1431.

Lindberg, L. & Hjern, A. (2003) Risk Factors for Anorexia Nervosa: A National Cohort Study. *International Journal of Eating Disorders,* 34(4): 397-408.

Lingswiler, V. M., Crowther, J. H. & Stephens, M. A. (1989) Affective and cognitive antecedents to eating episodes in bulimia and binge eating. *International Journal of Eating Disorders,* 8: 533-539.

Liu, Y., Gao, J. H., Liu, H. L. & Fox, P. T. (2000) The temporal response of the brain after eating revealed by functional MRI. *Nature,* 405: 1058-1062.

Lopez, C., Tchanturia, K., Stahl, D., Booth, R., Holliday, J. & Treasure, J. (2008) An examination of central coherence in women with Anorexia Nervosa. *International Journal of Eating Disorders,* 41(4): 340-347.

Lucas, A. R. (1981) Toward the understanding of anorexia nervosa as a disease entity. *Mayo Clinic Proceedings,* 56(4): 254-264; Cited in Katzman, D. K. & Golden, N. H. (2009) Anorexia Nervosa and Bulimia Nervosa, Chapter 33. In Neinstein, L. S., Gordon, C. M., Katzman, D. K., Rosen, D. S., Woods, E. R. (eds.) *Handbook of Adolescent Health Care.* Philadelphia: Lippincott, Williams and Wilkins, 321-332.

Mackenzie, R., Sims, C., Owens, R. G. & Dixon, A. K. (1995) Patients' perceptions of magnetic resonance imaging. *Clinical Radiology,* 50(3): 137-143.

Mahar, I., Bambico, F. R., Mechawar, N. & Nobrega, J. N. (2014) Stress, serotonin, and hippocampal neurogenesis in relation to depression and antidepressant effects. *Neuroscience & Biobehavioral Reviews,* 38: 173-192.

Mainz, V., Schulte-Ruether, M., Fink, G. R., Herpertz-Dahlmann, B. & Konrad, K. (2012) Structural brain abnormalities in adolescent anorexia nervosa before and after weight recovery and asscociated hormonal changes. *Psychosomatic Medicine,* 74(6): 574-582.

Makowski, A. C., Mnich, E. E., Angermeyer, M. C., Loewe, B. & von dem Knesebeck, O. (2014) Sex differences in attitudes toward females with eating disorders. *Eating Behavior,* 16C: 78-83.

Malkoff, S. B., Marcus, M. D., Grant, A., Moulton, M. M. & Vayonis, C. (1993) The relationship between dieting and binge eating among obese individuals. *Annual Behavioral Medicine,* 15: 40.

Marcus, M. D. & Wildes, J. E. (2009) Obesity: Is it a Mental Disorder? *International Journal of Eating Disorders,* 42(8): 739-753.

McCabe, C. & Rolls, E. T. (2007) Umami: a delicious flavour formed by convergence of taste and olfactory pathways in the human brain. *European Journal of Neurosciences,* 25(6): 1855-1864.

McCabe, C., Cowen, P. J. & Harmer, C. J. (2009) Neural representation of reward in recovered depressed patients. *Psychopharmacology,* 205: 667-677.

McCabe, C., Woffindale, C., Harmer, C. J. & Cowen, P. J. (2012) Neural processing of reward and punishment in young people at increased familial risk of depression. *Biological Psychiatry,* 72(7): 588-594.

McCarthy, M. (1990) The thin ideal, depression, and eating disorders in women. *Behavioral Research & Therapy,* 28: 205-218.

McEwen, B. S. & Gianaros, P. J. (2010) Central role of the brain in stress and adaption: Links to socioeconomic status, health, and disease. *Annals of the New York Academy of Sciences,* 1186: 190-222.

Mechelli, A., Price, C. J., Friston, K. J. & Ashburner, J. (2005) Voxel-Based Morphometry of the Human Brain: Methods and Applications. *Current Medical Images Reviews,* 1(2): 105-113.

Mertens, P., Blond, S., David, R. & Rigoard, P. (2015) Anatomy, physiology and neurobiology of nociception: A focus on low back pain. *Neurochirurgie,* 61(1): 22-34.

Milos, G., Spindler, A., Hepp, U. & Schnyder, U. (2004) Suicide attempts and suicidal ideation: Links with psychiatric co-morbidity in eating disorder subjects. *General hospital psychiatry,* 26(2): 129-135.

Milos, G., Spindler, A., Ruegsesser, P., Hasler, G., Schnyder, U., Laib, A., Gallo, L. M., Uebelhart, D. & Haeuselmann, H. (2007) Does weight gain induce cortical and trabecular bone regain in anorexia nervosa? A two-year prospective study. *Bone,* 41(5): 869-874.

Miwa, H., Nakanishi, I., Kodama, R. & Kondo, T. (2004) Cerebellar atrophy in a patient with anorexia nervosa. *International Journal of Eating Disorders,* 36: 238-241.

Miyake, Y., Okamoto, Y., Onoda, K., Kurosaki, M., Shirao, N., Okamoto, Y. & Yamawaki, S. (2010a) Brain activation during the perception of distorted body images. *Psychiatry Research,* 181: 183-192.

Miyake, Y., Okamoto, Y., Onoda, K., Shirao, N., Okamoto, Y., Otagaki, Y. & Yamawaki, S. (2010b) Neural processing of negative word stimuli concerning body image in patients with eating disorders: An fMRI study. *NeuroImage,* 50: 1333-1339.

Mohr, H. M., Zimmermann, J., Roeder, C., Lenz, C., Overbeck, G. & Grabhorn, R. (2010) Separating two components of body image in anorexia nervosa using fMRI. *Psychological Medicine,* 40: 1519-1529.

Morewedge, C. K., Huh, Y. E., & Vosgerau, J. (2010). Thought for food: imagined consumption reduces actual consumption. *Science,* 330: 1530–1533.

Morris, J. S. & Dolan, R. J. (2001) Involvement of human amygdala and orbitofrontal cortex in hunger-enhanced memory for food stimuli. *Journal of Neuroscience,* 21: 5304-5310.

Muehlau, M., Gaser, C., Ilg, R., Conrad, B., Leibl, C., Cebulla, M. H., Backmund, H., Gerlinghoff, M., Lommer, P., Schnebel, A., Wolfschlaeger, A. M., Zimmer, C. & Nunnemann, S. (2007) Grey matter decrease of the anterior cingulated cortex in anorexia nervosa. *American Journal of Psychiatry,* 164: 1850-1857.

Nagai, M., Kishi, K. & Kato, S. (2007) Insula cortex and neuropsychiatric disorders: A review of recent literature. *European Psychiatry,* 22: 387-394.

Nakahara, R., Yoshiuchi, K., Yamanaka, G., Sasaki, T., Suematsu, H. & Kuboki, T. (2000) Coping skills in Japanese women with eating disorders. *Psychological Reports,* 87: 741-746.

Nakamua, Y., Goto, T. K., Tokumori, K., Yoshiura, T., Kobayashi, K., Nakamura, Y., Honda, H., Ninomiya, Y. & Yoshiura, K. (2011) Localization of brain activation by umami taste in humans. *Brain Research,* 1406: 18-29.

Naruo, T., Nakabeppu, Y., Sagiyama, K., Munemoto, T., Homan, N., Deguchi, D., Nakajo, M. & Nozoe, S. (2000) Characteristic regional cerebral blood flow patterns in anorexia nervosa patients with binge/purge behavior. *American Journal of Psychiatry,* 157: 1520-1522.

Nikendei, C., Funiok, C., Pfueller, U., Aschenbrenner, S., Weisbrod, M., Herzog, W. & Friedrich, H. C. (2010). Memory performance in acute and weight-restored anorexia nervosa patients. *Psychological Medicine,* 9: 1-10.

Nunn, K., Frampton, I., Gordon, I. & Lask, B. (2008) The fault is not in her parents but in her insula - a neurobiological hypothesis of anorexia nervosa. *European Eating Disorder Review,* 16(5): 355-60.

Nunn, K., Lask, B. & Frampton, I. (2011) Towards a comprehensive, causal and explanatory neuroscience model of anorexia nervosa. In Lask, B. & Frampton, I. (eds.) *Eating Disorders and the Brain.* Chichester: Wiley-Blackwell, 164-179.

Oberndorfer, T. A., Frank, G. A., Simmons, A. N., Wagner, A., McCurdy, D., Fudge, J. L., Yang, T. T., Paulus, M. P. & Kaye, W. H. (2013) Altered insula response to sweet taste processing after recovery from anorexia and bulimia nervosa. *American Journal of Psychiatry,* 170(10): 1143-1151.

O'Doherty, J., Rolls, E. T., Francis, S., Bowtell, R. & McGlone, F. (2001) Representation of Pleasant and Aversive Taste in Human Brain. *Journal of Neurophysiology,* 85: 1315-1321.

O'Doherty, J. P., Deichman, R., Critchley, H. D. & Dolan, R. J. (2002) Neural responses during anticipation of a primary taste reward. *Neuron,* 33: 815-826.

O'Doherty, J. P. (2004) Reward representations and reward-related learning in the human brain: insights from neuroimaging. *Current Opinion in Neurobiology,* 14: 769-776.

O'Doherty, J. P. (2007) Lights, camembert, action! The role of human orbitofrontal cortex in encoding stimuli, rewards, and choices. *Annals of the New York Academy of Sciences,* 1121: 254-272.

Papagni, S. A., Benetti, S., Arulanantham, S., McCrory, E., McGuire, P. & Mechelli, A. (2011) Effects of stressful life events on human brain structure: a longitudinal voxel-based morphometry study. *Stress,* 14(2): 227-232.

Paul, T. & Thiel, A. (2005) *EDI-2. Eating Disorder Inventory-2.* Deutsche Version. Goettingen: Hogrefe.

Peat, C., Mitchell, J. E., Hoek, H. W. & Wonderlich, S. A. (2009) Validity and Utility of Subtyping Anorexia Nervosa. *International Journal of Eating Disorders,* 42(7): 590-594.

Pelchat, M. L., Johnson, A., Chan, R., Valdez, J. & Ragland, J. D. (2004) Images of desire: Food-craving activation during fMRI. *NeuroImage,* 23: 815-826.

Perpina, C., Hemsley, D., Treasure, J. & de Silva, P. (1993) Is the selective information processing of food and body words specific to patients with eating disorders? *International Journal of Eating Disorders,* 14: 359-366.

Phillips, M. L., Drevets, W. C., Rauch, S. L. & Lane, R. (2003) Neurobiology of emotion perception I: The neural basis of normal emtion perception. *Biological Psychiatry,* 54(5): 504.514.

Pietrowsky, R., Krug, R., Fehm, H. L. & Born, J. (2002) Food deprivation fails to affect preoccupation with thoughts of food in anorectic patients. *British Journal of Clinical Psychology,* 41: 321-326.

Polivy, J. & Herman, C. P. (2002) Causes of Eating Disorders. *Annual Review of Psychology,* 53: 187-213.

Porubska, K. L., Veit, R., Preissl, H., Fritsche, A. & Birbaumer, N. (2006) Subjective feeling of appetite modulates brain activity. An fMRI study. *NeuroImage,* 32: 1273-1280.

Pruis, T. A., Keel, P. K. & Janowsky, J. S. (2012) Recovery from anorexia nervosa includes neural compensation for negative body image. *International Journal of Eating Disorders,* 45(8): 919-931.

Pryor, T., Wiederman, M. W. & McGilley, B. (1996) Clinical correlates of anorexia nervosa subtypes. *International Journal of Eating Disorders,* 19: 371-379.

Radeloff, D., Willmann, K., Otto, L., Lindner, M., Putnam, K., Leeuwen, S. V., Kaye, W. H., Poustka, F. & Wagner, A. (2014) High-fat taste challenge reveals altered striatal response in women recovered from bulimia nervosa: A pilot study. *World Journal of Biological Psychiatry,* 15(4):307-316.

Reba, L., Thornton, L., Tozzi, F., Klump, K. L., Brandt, H., Crawford, S., Crow, S., Fichter, M. M., Halmi, K. A., Johnson, C., Kaplan, A. S., Keel, P., LaVia, M., Mitchell, J., Strober, M., Woodside, D. B., Rotondo, A., Berrettini, W. H., Kaye, W. H. & Bulik, C. M. (2005) Relationship between features associated with vomiting in purging-type eating disorders. *International Journal of Eating Disorders,* 38(4): 287-294.

Rimol, L. M., Hartberg, C. B., Nesvag, R., Fennema-Notestine, C., Hagler, D. J., Pung, C. J., Jennings, R. G., Haukvig, U. K., Lange, E., Nakstad, P. H., Melle, I., Andreassen, O. A., Dale, A. M. & Agartz, I. (2010) Cortical thickness and subcortical volumes in schizophrenia and bipolar disorder. *Biolocial Psychiatry,* 68(1): 41-50.

Roberts, M. E., Tchanturia, K. & Treasure, J. L. (2010) Exploring the neurocognitive signature of poor set-shifting in anorexia and bulimia nervosa. *Journal of Psychiatric Research,* 44: 964-970.

Robertson, P. (2009) *Severe and enduring eating disorder (SEED).* Wiley-Blackwell.

Rocha-Rego, V., Pereira, M. G., Oliveira, L., Mendlowicz, M. V., Fiszman, A., Marques-Portella, C., Berger, W., Chu, C., Joffily, M., Moll, J.,

Mari, J. J., Figueira, I. & Volchan, E. (2012) Decreased premotor cortex volume in victims of urban violence with posttraumatic stress disorder. *PLoS One,* 7(8): e42560.

Rodrigues, E., Wenzel, A., Ribeiro, M. P., Quarantini, L. C., Miranda-Scippa, A., de Sena, E. P. & de Oliveira, I. R. (2011) Hippocampal volume in borderline personality disorder, with and without comorbid posttraumatic stress disorder: a meta-analysis. *European Psychiatry,* 26(7): 452-456.

Rudenga, K., Green, B., Nachtigal, D. & Small, D. M. (2010) Evidence for an integrated oral sensory module in the human anterior ventral insula. *Chemical Senses,* 35(8): 693-703.

Roefs, A., Herman, C. P., MacLeod, C. M., Smulders, F. T. & Jansen, A. (2005) At first sight: how do restrained eaters evaluate high-fat palatable foods? *Appetite,* 44: 103-114.

Rothemund, Y., Preuschhof, C., Bohner, G., Bauknecht, H. C., Klingebiel, R., Flor, H. & Klapp, B. F. (2007) Differential activation of the dorsal striatum by high-caloric visual food stimuli in obese individuals. *NeuroImage,* 37: 410-421.

Sachdev, P., Mondraty, N., Wen, W. & Gulliford. K. (2008) Brains of anorexia nervosa patients process self-images differently from non-self images: An fMRI study. *Neuropsychologia,* 46: 2161-2168.

Salat, D. H., Buckner, R. L., Snyder, A. Z., Greve, D. N., Desikan, R. S. R., Busa, E., Morris, J. C., Dale, A. M. & Fischl, B. (2004) Thinning of the cerebral cortex in ageing. *Cerebral Cortex,* 14(7): 721-730.

Santel, S., Baving, L., Krauel, K., Muente, T. F. & Rotte, M. (2006) Hunger and satiety in anorexia nervosa: fMRI during cognitive processing of food pictures. *Brain Research,* 1114: 138-148.

Saper, C. B., Chou, T. C. & Elmquist, J. K. (2002) The need to feed: homeostatic and hedonic control of eating. *Neuron,* 36: 199-211.

Schienle, A., Schaefer, A., Hermann, A. & Vaitl, D. (2009) Binge-Eating Disorder: Reward sensivity and brain activation to images of food. *Biological Psychiatry,* 65: 654-661.

Schwartz, M. W., Woods, S. C., Porte, D., Seeley, R. J. & Baskin, D. G. (2000) Central nervous system control of food intake. *Nature,* 404: 661-671.

Seed, J. A., Dixon, R. A., McCluskey, S. E. & Young, A. H. (2000) Basal activity of the hypothalamic-pituitary-adrenal axis and cognitive function in Anorexia Nervosa. *European Archives of Psychiatry and Clinical Neuroscience,* 250(1): 11-15.

Seeger, G., Braus, D. F., Ruf, M., Goldberger, U. & Schmidt, M. H. (2002) Body image distortion reveals amygdala activation in patients with anorexia nervosa – a functional magnetic resonance imaging study. *Neuroscience Letters,* 326: 25-28.

Simonnet, J., Eugène, E., Cohen, I., Miles, R. & Fricker, D. (2013) Cellular neuroanatomy of rat presubiculum. *European Journal of Neuroscience,* 37(4): 583-597.

Simmons, W. K., Martin, A. & Barsalou, L. W. (2005) Pictures of appetizing foods activate gustatory cortices for taste and reward. *Cerebral Cortex,* 15: 1602-1608.

Simmons, W. K. & Drevets, W. C. (2012) A "taste" of what is to come: reward sensivbity as a potential endophenotype for major depressive disorder. *Biological Psychiatry,* 72(7):526-527.

Small, D. M., Zatorre, R. J., Dagher, A., Evans, A. C. & Jones-Gotman, M. (2001) Changes in brain activity related to eating chocolate: From pleasure to aversion. *Brain,* 124: 1720-1733.

Small, D. M., Jones-Gotman, M. & Dagher, A. (2003) Feeding-induced dopamine release in dorsal striatum correlates with meal pleasantness ratings in healthy human volunteers. *NeuroImage,* 19: 1709-1715.

Small, D. M., Bender, G., Veldhuizen, M. G., Rudenga, K., Nachtigal, D. & Felsted, J. (2007) The role of the human orbitofrontal cortex in taste and flavor processing. *Annals of the New York Academy of Sciences,* 1121: 136-151.

Smeets, P. A. M., de Graaf, C., Stafleu, A., van Osch, M. J. P., Nievelstein, R. A. J. & van der Grond, J. (2006) Effect of satiety on brain activation during chocolate tasting in men and women. *American Journal of Clinical Nutrition,* 83: 1297-1305

Sowell, E. R., Thompson, P. M., Leonard, C. M., Welcome, S. E., Kan, E. & Toga, A. W. (2004) Longitudinal mapping of cortical thickness and brain growth in normal children. *The Journal of Neuroscience,* 24(38): 8223-8231.

Spetter, M. S., Smeets, P. A., de Graaf, C. & Viergever, M. A. (2010) Representation of sweet and salty taste intensity in the brain. *Chemical Senses,* 35(9): 831-840.

Steinhausen, H. C. (2002) The Outcome of Anorexia Nervosa in the 20th Century. *American Journal of Psychiatry,* 159(8): 1284-1293.

Stoeckel, L. E., Cox, J. E., Cook, E. W. & Weller, R. E. (2007) Motivational state modulates the hedonic value of food images differently in men and women. *Appetite,* 48: 139-144.

St Onge, M. P., Sy, M., Heymsfield, S. B. & Hirsch, J. (2005) Human cortical specialization for food: a functional magnetic resonance imaging investigation. *Journal of Nutrition,* 135: 1014-1018.

Strauß, B. & Richter-Appelt, H. (1996) FBeK. *Fragebogen zur Beurteilung des eigenen Körpers.* Goettingen: Hogrefe.

Strober, M., Freeman, R., Lampert, C., Diamond, J. & Kaye, W. (2000) Controlled Family Study of Anorexia Nervosa and Bulimia Nervosa: Evidence of Shared Liability and Transmission of Partial Syndromes. *American Journal of Psychiatry,* 157(3): 393-401.

Stuss, D. T. & Benson, D. F. (1986) *The Frontal Lobes.* New York: Raven.

Sullivan, P. F. (1995) Mortality in anorexia nervosa. *American Journal of Psychiatry,* 15: 1073-1074.

Swanson, H., Power, K., Collin, P., Deas, S., Paterson, G., Grierson, D., Yellowlees, A., Park, K. & Taylor, L. (2010) The relationship between parental bonding, social problem solving and eating pathology in an anorexic inpatient sample. *European Eating Disorder Reviews,* 18(1): 22-32.

Takano, A., Shiga, T., Kitagawa, N., Koyama, T., Katoh, C., Tsukamoto, E. & Tamaki, N. (2001) Abnormal neuronal network in anorexia nervosa studied with I-123-IMP SPECT. *Psychiatry Research,* 107: 45-50.

Tataranni, P. A. & Delparigi, A. (2003) Functional neuroimaging: a new generation of human brain studies in obesity research. *Obesity Reviews*, 4: 229-238.

Taylor, S. E. & Stanton, A. L. (2007) Coping resources, coping processes, and mental health. *Annual Review of Clinical Psychology*, 3: 377-401.

Tchanchuria, K., Morris, R. G., Anderluh, M. B., Collier, D. A., Nikolaou, V. & Treasure, J. (2004) Set shifting in anorexia nervosa: An examination before and after weight gain, in full recovery and relationship to childhood and adult OCPD traits. *Journal of Psychiatric Research*, 38: 545-552.

Tejado, L., de Anta Ruiz, R. M., Trebbau, H., Diaz-Marsá, M. & Perera, J. L. C. (2010) Functional magnetic resonance studies in eating behaviour disorders. *Actas Espanolas de Psiquiatria*, 38: 183-188.

Tencono, E., Santonastaso, P., Degortes, D., Bosello, R., Titton, F., Mapelli, D. & Favaro, A. (2010) Set-shifting abilities, central coherence, and handedness in anorexia nervosa patients, their unaffected siblings and healthy controls. Exploring Putative Endophenotypes. *The World Journal of Biological Psychiatry*, 11: 813-82.

Thornton, L. M., Mazzeo, S. E. & Bulik, C. M. (2011) The heritability of eating disorders: methods and current findings. In Adan, R. A. H. & Kaye, W. H. (eds.) *Behavioral neurobiology of eating disorders*. Berlin, Heidelberg: Springer Verlag, 141-156.

Tozzi, F., Thornton, L. M., Klump, K. L., Fichter, M. M., Halmi, K. A., Kaplan, A. S., Strober, M., Woodside, D. B., Crow, S., Mitchell, J., Rotondo, A., Mauri, M., Cassano, G., Keel, P., Plotnicov, K. H., Pollice, C., Lilenfeld, L. R., Berrettini, W. H., Bulik, C. M. & Kaye, W. H. (2005) Symptom fluctuation in eating disorders: correlates of diagnostic crossover. *American Journal of Psychiatry*, 162(4): 732-740.

Tozzi, F., Thornton, L. M., Mitchell, J., Fichter, M. M., Klump, K. L., Lilenfeld, L. R., Reba, L., Strober, M., Kaye, W. H., Bulik, C. M. & Price Foundation Collaborative Group (2006) Features associated with laxative abuse in individuals with eating disorders. *Psychosomatic Medicine*, 68: 470-477.

Treasure, J. & Holland, A. (1995) Genetic factors in eating disorders. In Szmukler, G., Dare, C. & Treasure, J. (eds.) *Handbook of Eating Disorders: Theory, Treatment and Research.* Wiley: Chichester, 65-81.

Troop, N. A., Holbrey, A. & Treasure, J. L. (1998) Stress, coping, and crisis support in eating disorders. *International Journal of Eating Disorders,* 24(2): 157-166.

Uher, R., Brammer, M. J., Murphy, T., Campbell, I. C., Ng, V. W., Williams, S. C. R. & Treasure, J. (2003) Recovery and Chronicity in Anorexia Nervosa: Brain Activity Associated with Differential Outcomes. *Biological Psychiatry,* 54: 934-942.

Uher, R., Murphy, T., Brammer, M. J., Dalgleish, T., Phillips, M. L., Ng, V. W., Andrew, C. M., Williams, S. C., Campbell, I. C. & Treasure, J. (2004) Medial prefrontal cortex activity associated with symptom provocation in eating disorders. *American Journal of Psychiatry,* 161: 1238–1246.

Uher, R., Murphy, T., Friederich, H. C., Dalgleish, T., Brammer, M. J., Giampietro, V., Giampetro, V., Phillips, M. L., Andrew, C. M., Ng, V. W., Williams, S. C., Campbell, I. C. & Treasure, J. (2005) Functional neuroanatomy of body shape perception in healthy and eating-disordered women. *Biological Psychiatry,* 58: 990-997.

Uher, R., Treasure, J., Heining, M., Brammer, M. J. & Campbell, I. C. (2006) Cerebral processing of food related stimuli: Effects of fasting and gender. *Behavioral Brain Research,* 169: 111-119.

Van Erp, T. G., Greve, D. N., Rasmussen, J., Turner, J., Calhoun, V. D., Young, S., Mueller, B., Brown, G. G., McCarthy, G., Glover, G. H., Lim, K. O., Bustillo, J. R., Belger, A., McEwen, S., Voyvodic, J., Mathalon, D. H., Keator, D., Preda, A., Nguyen, D., Ford, J. M., Potkin, S. G. & FBIRN. (2014) A multi-scanner study of subcortical brain volume abnormalities in schizophrenia. *Psychiatry Research,* 222(1-2): 10-16.

Van Leemput, K., Bakkour, A., Benner, T., Wiggins, G., Wald, L. L., Augustinack, J., Dickerson, B. C., Golland, P. & Fischl, B. (2009) Automated segmentation of hippocampal subfields from ultra-high resolution in vivo MRI. *Hippocampus,* 19(6): 549-557.

Van Rijn, I., de Graaf, C. & Smeets, P. A. (2014) Tasting calories differentially affects brain activation during hunger and satiety. *Behavioral Brain Research,* 279C: 139-147.

Vaz, F. J., Alcaina, T. & Guisado, J. A. (1998) Food aversions in eating disorders. *International Journal of Food Sciences and Nutrition,* 49: 181-186.

Veldhuizen, M. G., Nachtigal, D., Teulings, L., Gitelman, D. R. & Small, D. M. (2010) The insula taste cortex contributes to odor quality coding. *Frontiers in Human Neuroscience,* 4: 58.

Vermetten, E., Schmahl, C., Lindner, S., Loewenstein, R. J. & Bremner, J. D. (2006) Hippocampal and amygdalar volumes in dissociative identity disorder. *American Journal of Psychiatry,* 163(4): 630-636.

Villa, V., Manzoni, G. M., Pagnini, F., Castelnuovo, G., Cesa, G. L. & Molinari, E. (2009) Do coping strategies discriminate eating disordered individuals better than eating disorder features? An explorative study on female inpatients with anorexia and bulimia nervosa. *Journal of Clinical Psychology in Medical Settings,* 16(4): 297-303.

Vocks, S., Busch, M., Schulte, D., Groenemeyer, D., Herpertz, S. & Suchan, B. (2010) Effects of body image therapy on the activation of the extrastriate body area in anorexia nervosa: an fMRI study. *Psychiatry Research,* 183: 114-118.

Vocks, S., Herpertz, S., Rosenberger, C., Senf, W. & Gizewski, E. R. (2011) Effects of gustatory stimulation on brain activity during hunger and satiety in females with restricting-type anorexia nervosa: an fMRI study. *Journal of Psychiatric Research,* 45: 395-403.

Volkow, N. D., Wang, G. J., Maynard, L., Jayne, M., Fowler, J. S., Zhu, W., Logan, J., Gatley, S. J., Ding, Y. S., Wong, C. & Pappas, N. (2003) Brain dopamine is associated with eating behavior in humans. *International Journal of Eating Disorders,* 33: 136-142.

Volkow, N. D. & Wise, R. A. (2005) How can drug addiction help us understand obesity? *Nature Neuroscience,* 8: 555-560.

Wagner, A., Greer, P., Bailer, U. F., Frank, G. K., Henry, S. E., Putnam, K., Meltzer, C. C., Ziolko, S. K., Hoge, J., McConaha, C. & Kaye, W. H. (2006a) Normal brain tissue volumes after long-term recovery in anorexia and bulimia nervosa. *Biological Psychiatry,* 59: 291-293.

Wagner, A., Aizenstein, H., Frank, G. K., Figurski, J., May, J. C., Putnam, K., Fischer, L., Bailer, U. F., Henry, S. E., McConaha, C., Vogel, V. & Kaye, W. H. (2006b) Neural correlates of habituation to taste stimuli in healthy women. *Psychiatry Research: Neuroimaging,* 147: 57-67.

Wagner, A., Aizenstein, H., Venkatraman, V. K., Fudge, J., May, J. C., Mazurkewicz, L., Frank, G. K., Bailer, U. F., Fischer, L., Nguyen, V., Carter, C., Putnam, K. & Kaye, W. H. (2007) Altered reward processing in women recovered form anorexia nervosa. *American Journal of Psychiatry,* 164: 1850-1857.

Wagner, A., Aizenstein, H., Mazurkewicz, L., Fudge, J., Frank, G. K., Putnam, K., Bailer, U. F., Fischer, L. & Kaye, W. H. (2008) Altered insula response to taste in stimuli in individuals recovered from restrictive-type Anorexia Nervosa. *Neuropsychopharmacology,* 33: 513-523.

Wang, G. J., Volkow, N. D., Telang, F., Jayne, M., Ma, J., Rao, M., Zhu, W., Wong, C. T., Pappas, N. R., Geliebter, A. & Fowler, J. S. (2004) Exposure to appetitive food stimuli markedly activates the human brain. *NeuroImage,* 21: 1790-1797.

Wang, Z., Neylan, T. C., Mueller, S. G., Lenoci, M., Truran, D., Marmar, C. R., Weiner, M. W. & Schuff, N. (2010) Magnetic resonance imaging of hippocampal subfields in posttraumatic stress disorder. *Archives of General Psychiatry,* 67(3): 296-303.

Watkins, B. (2011) Eating disorders: an overview. In Lask, B. & Frampton, I. *Eating Disorders and the Brain.* Oxford, Chichester, New York: Wiley-Blackwell, 19-55.

Weider, S., Indredavik, M. S., Lydersen, S. & Hestad, K. (2015) Neuropsychological functioning in patients with Anorexia Nervosa or bulimia nervosa. *International Journal of Eating Disorders,* 48(4): 397-405.

Wilson, G. T., Becker, C. B. & Heffernan, K. (2003) Eating disorders. In Barkley, R. A. & Mash, E. J. (eds.) *Child Psychopathology.* New York: The Guilford Press, 687-715.

Wittchen, H. U., Zaudig, M., Fydrich, T. (1997) Strukturiertes klinisches Interview für DSM-IV. Goettingen: Hogrefe.

Wood, D. (2011) Why clinicians should love neuroscience: the clinical relevance of contemporary knowledge. In Lask, B. & Frampton, I. (eds.) *Eating Disorders and the Brain.* Oxford, Chichester, New York: Wiley-Blackwell, 1-18.

World Health Organization (1994) *The ICD-10. Classification of Mental and Behavioural Disorders.* Available at: http://www.who.int/classifications/icd/en/bluebook.pdf?ua=1 [accessed on December 04, 2014].

World Health Organization (1997) *Obesity. Preventing and Managing the Global Epidemic.* Geneva:.Report of a WHO Consultation on Obesity.

World Health Organization (1996) *WHOQOL-BREF. Introduction, Administration, Scoring, and Generic Version of the Assessment.* Geneva: WHO. http://www.who.int/mental_health/media/en/76.pdf [accessed on December 04, 2014].

Youssef, G., Plancherel, B., Laget, J., Corcos, M., Flament, M. F. & Halfon, O. (2004) Personality trait risk factors for attempted suicide among young women with eating disorders. *European Psychiatry,* 19: 131-139.

Zakzanis, K. K., Campbell, Z. & Polsinelli, A. (2010) Quantitative evidence for distinct cognitive impairment in anorexia nervosa and bulimia nervosa. *Journal of Neuropsychology,* 4(1): 89-106.

Zald, D. H. & Pardo, J. V. (2000) Cortical activation induced by intraoral stimulation with water in humans. *Chemical Senses,* 25: 267-275.

Zald, D. H. & Kim, S. W. (1996) Anatomy and function of the orbital frontal cortex I: anatomy, neurocircuitry, and obsessive-compulsive disorder. *Journal of Neuropsychiatry and Clinical Neurosciences,* 8(2): 125-138.

6 Appendix

Table 23: Test statistics concerning differences in brain activity due to the administration of sucrose between AN patients and CW in the ROIs

x	y	z	pixel	Area	AN Mean	CW Mean	p
45	-4	1	19	insula	-0.22	-0.77	**.030**
-12	29	16	32	ACC	-0.12	-0.35	.052
3	-52	7	71	ACC	-0.49	-1.19	.080
9	-46	4	71	ACC	-0.45	-1.13	**.048**
0	-10	-20	16	ACC	-0.55	0.79	**.008**
-24	32	-11	100	frontal cortex	-0.13	-0.42	**.020**
-12	47	-8	100	FG	0.01	-0.47	**.022**
-21	47	-8	100	middle FG	0.03	-0.27	.094
27	47	-8	13	middle FG	0.21	-0.14	**.027**
18	50	19	15	superior FG	-0.07	-0.41	**.025**
42	-34	25	7	IPL	-0.19	-0.45	.086
60	-28	40	14	BA 1	-0.15	-0.53	.110
-21	-1	70	23	BA 6	-0.47	-0.85	.118
3	-61	55	74	BA 7	-0.13	-0.50	.097
-27	50	40	9	BA 9	-0.74	-0.15	**.014**
-6	-61	67	74	precuneus left	-0.11	-0.83	.076
3	-43	67	74	postcentral gyrus	-0.34	-0.99	.073
3	-1	4	34	lateral ventricle	-0.33	-1.07	.052
9	-1	-14	16	right cerebrum	-0.69	0.13	**.047**
-12	32	7	32	corpus callosum	-0.04	-0.29	**.017**
-3	26	4	32	corpus callosum	-0.19	-0.47	**.047**
-6	-37	4	6	corpus callosum	-0.30	-0.77	.212

x	y	z	pixel	Area	AN Mean	CW Mean	p
21	-13	-14	23	PHG	-0.22	-1.07	.068
24	5	-14	5	BA 34	-0.26	-0.78	.097

Results of MANOVA. The significant p-values are bolded. AN: women with anorexia nervosa, restrictive type (N=20), CW: healthy control women (N=20), SD: standard deviation, ACC: anterior cingulate cortex, FG: frontal gyrus, IPL: inferior parietal lobule, BA: Brodmann area, PHG: parahippocampal gyrus.

Table 24: Test statistics concerning differences in brain activity due to the administration of umami between AN patients and CW in the ROIs

x	y	z	pixel	Area	AN Mean	CW Mean	p
45	-4	1	19	insula	-0.21	-0.81	**.014**
-12	29	16	32	ACC	-0.12	-0.36	**.047**
3	-52	7	71	ACC	-0.33	-1.21	**.023**
9	-46	4	71	ACC	-0.28	-1.10	**.014**
0	-10	-20	16	ACC	-0.50	0.63	**.014**
-24	32	-11	100	frontal cortex	-0.16	-0.47	**.019**
-12	47	-8	100	FG	0.02	-0.49	**.006**
-21	47	-8	100	middle FG	-0.02	-0.34	.051
27	47	-8	13	middle FG	0.13	-0.18	**.048**
18	50	19	15	superior FG	-0.07	-0.40	**.033**
42	-34	25	7	IPL	-0.19	-0.49	**.041**
60	-28	40	14	BA 1	-0.16	-0.52	.115
-21	-1	70	23	BA 6	-0.46	-0.91	.072
3	-61	55	74	BA 7	-0.13	-0.47	.094
-27	50	40	9	BA 9	-0.67	-0.19	.061
-6	-61	67	74	precuneus left	-0.24	-0.90	.083
3	-43	67	74	postcentral gyrus	-0.50	-1.02	.117
3	-1	4	34	lateral ventricle	-0.35	-1.11	**.041**
9	-1	-14	16	right cerebrum	-0.60	0.02	.097
-12	32	7	32	corpus callosum	-0.04	-0.31	**.009**
-3	26	4	32	corpus callosum	-0.17	-0.48	**.027**
-6	-37	4	6	corpus callosum	-0.23	-0.88	.067
21	-13	-14	23	PHG	-0.16	-1.05	**.043**
24	5	-14	5	BA 34	-0.28	-0.80	.102

Results of MANOVA. The significant p-values are bolded. AN: women with anorexia nervosa, restrictive type (N=20), CW: healthy control women (N=20), SD: standard deviation, ACC: anterior cingulate cortex, FG: frontal gyrus, IPL: inferior parietal lobule, BA: Brodmann area, PHG: parahippocampal gyrus.

Table 25: Test statistics concerning differences in brain activity due to the administration of citric acid between AN patients and CW in the ROIs

x	y	z	pixel	Area	AN Mean	CW Mean	p
45	-4	1	19	insula	-0.12	-0.77	**.008**
-12	29	16	32	ACC	-0.11	-0.31	.129
3	-52	7	71	ACC	-0.36	-1.14	**.034**
9	-46	4	71	ACC	-0.35	-1.07	**.037**
0	-10	-20	16	ACC	-0.75	0.70	**.002**
-24	32	-11	100	frontal cortex	-0.12	-0.39	**.033**
-12	47	-8	100	FG	-0.08	-0.41	.094
-21	47	-8	100	middle FG	-0.09	-0.21	.518
27	47	-8	13	middle FG	0.04	-0.10	.386
18	50	19	15	superior FG	-0.09	-0.36	.127
42	-34	25	7	IPL	-0.16	-0.47	**.037**
60	-28	40	14	BA 1	-0.06	-0.51	.057
-21	-1	70	23	BA 6	-0.48	-0.81	.139
3	-61	55	74	BA 7	-0.06	-0.45	.096
-27	50	40	9	BA 9	-0.66	-0.18	.052
-6	-61	67	74	precuneus left	-0.11	-0.85	.055
3	-43	67	74	postcentral gyrus	-0.39	-0.98	.076
3	-1	4	34	lateral ventricle	-0.27	-1.10	**.020**
9	-1	-14	16	right cerebrum	-0.85	0.13	**.020**
-12	32	7	32	corpus callosum	-0.06	-0.27	**.040**
-3	26	4	32	corpus callosum	-0.19	-0.45	**.037**
-6	-37	4	6	corpus callosum	-0.29	-0.80	.163
21	-13	-14	23	PHG	-0.12	-1.03	**.043**
24	5	-14	5	BA 34	-0.14	-0.76	**.048**

Results of MANOVA. The significant p-values are bolded. AN: women with anorexia nervosa, restrictive type (N=20), CW: healthy control women (N=20), SD: standard deviation, ACC: anterior cingulate cortex, FG: frontal gyrus, IPL: inferior parietal lobule, BA: Brodmann area, PHG: parahippocampal gyrus.

Publications and presentations resulting from this dissertation

Publication:

Burkert, N. T., Koschutnig, K., Ebner, F. & Freidl, W. (2015) Structural hippocampal alterations, perceived stress, and coping deficiencies in patients with anorexia nervosa. *International Journal of Eating Disorders*, epub ahead of print. DOI: 10.1002/eat.22397.

Presentations at scientific events:

Burkert, N. T., Koschutnig, K., Ebner, F. & Freidl, W. (2014) Brain activity in response to different taste stimuli in subjects suffering from anorexia nervosa compared to healthy controls. International Conference on Eating Disorders, MARCH 27-29, 2014, New York, USA. 2014. [Poster]

Burkert, N. T., Koschutnig, K., Ebner, F. & Freidl, W. (2014) Structural hippocampal alterations, perceived stress, and coping deficiencies in Anorexia Nervosa. International Conference on Eating Disorders, MARCH 27-29, 2014, New York, USA. 2014. [Poster]

Burkert, N. T., Koschutnig, K., Ebner, F. & Freidl, W. (2014) Brain activity in response to different taste stimuli in subjects suffering from anorexia nervosa compared to healthy controls. 4th scientific conference of the German Society of Eating Disorders, MARCH 20-22, 2014, Leipzig, Germany. 2014. [Oral Communication]

Burkert, N. T., Koschutnig, K., Ebner, F. & Freidl, W. (2014) Hippocampal volume changes, perceived stress, and coping deficiencies in Anorexia Nervosa. 4th scientific conference of the German Society of Eating Disorders, MARCH 20-22, 2014, Leipzig, Germany. 2014. [Poster]

Burkert, N. T., Koschutnig, K., Ebner, F. & Freidl, W. (2013) Hippocampal volume reduction, perceived stress, and coping deficiencies in Anorexia Nervosa. 21st international conference on Eating Disorders, OCT 17-19, 2013, Alpbach, Austria. 2013. [Oral Communication]

Burkert, N. T., Koschutnig, K., Ebner, F. & Freidl, W. (2013) Brain activity in response to different taste stimuli in subjects suffering from Anorexia Nervosa in comparison to healthy controls. XXI World Congress on Neurology: September 21 – 26, 2013; Vienna, AUSTRIA. 2013. [Poster]

Burkert, N. T., Koschutnig, K., Ebner, F. & Freidl, W. (2013) Hippocampal changes and associations with psychological symptoms in subjects suffering from Anorexia Nervosa. XXI World Congress on Neurology: September 21 – 26, 2013; Vienna, AUSTRIA. 2013. [Poster]

Burkert, N. T., Koschutnig, K., Ebner, F. & Freidl, W. (2013) Hippocampal changes and associations with stress symptoms in subjects suffering from Anorexia Nervosa. European Council on Eating Disorders: September 12 – 14, 2013; Oslo, NORWAY. 2013. [Oral Communication]

MIX
Papier aus verantwortungsvollen Quellen
Paper from responsible sources
FSC® C105338

If you have any concerns about our products,
you can contact us on
ProductSafety@springernature.com

In case Publisher is established outside the EU,
the EU authorized representative is:
Springer Nature Customer Service Center GmbH
Europaplatz 3, 69115 Heidelberg, Germany

Printed by Libri Plureos GmbH
in Hamburg, Germany